PRAISE FOR GEORGE MANSOUR

"I can't say enough about George Mansour's knowledge about this area! His approach in the book is to keep it simple and not make it too complicated to protect your personal and business data. Before it came to our front door, he was preaching about the need for data security and the necessary steps that businesses, small and large, need to undergo to not only protect their clients, but themselves!

His insights have helped my firm and my clients. He pinpointed within five minutes what took many weeks for forensics to confirm. Read the book, take his suggestions, and implement them. Be prepared!"

—**Pam,**
Attorney

"I've learned how important it is to not make quick decisions and not be in a rush to do things that will affect our computer network. Every time I have acted without confirming my actions with George, it has caused everyone aggravation, extra work and risk to the network. I value George's knowledge and continuing education, and I know my network is safe!"

—Linda,
Entrepreneur and Business Owner

"George gives us the ability to handle all problems...no issue is too large or hard to overcome. More importantly, George is always ahead of the curve in an industry that changes by the minute and creates more issues and obstacles for systems and security."

—Jeffrey,
Attorney

UNHACKABLE

YOUR ONLINE SECURITY PLAYBOOK
RECREATING CYBERSECURITY IN AN
UNSECURE WORLD

GEORGE MANSOUR

LASTING PRESS

ISBN: 978-1-949696-04-2 (mobi)
ISBN: 978-1-949696-05-9 (epub)
ISBN: 978-1-949696-06-6 (paperback)
Printed in the United States of America

Published by:

Lasting Press
615 NW 2nd Ave. #915
Canby, OR 97013

Cover and Interior Design by: Rory Carruthers Marketing

Project Management and Book Launch by: Rory Carruthers Marketing
www.RoryCarruthers.com

For more information about George Mansour or to book him for your next event, speaking engagement, podcast or media interview, please visit: www.GeorgeMansour.com

DEDICATED TO HUMANITY AND FUTURE GENERATIONS

Special thanks to my family and friends, whose love, strength, and support allowed me to follow my dreams.

All Is Grace

CONTENTS

FOREWORD

One typically cold December Sunday morning in New England, my family and I were not in the mood to cook breakfast. A better idea emerged when we decided to eat out at the local breakfast restaurant. The waiting line that morning was extensive and out the door. My wife approached the hostess station and was informed the wait was about thirty minutes. The question came into play: do we wait or go somewhere else?

Our children were small at the time, and the idea of packing them up to leave and drive somewhere else did not seem prudent. The waiting area was packed with other patrons since the weather forecast showed a chilly 32 degrees. Suddenly, we lucked out and got a seat on the bench to sit while we waited.

I have a background in criminal justice with over twenty years of public service in law enforcement and corrections. Most of my time was spent in the parole/probation field, but later I segued into the private sector as an executive protection operator. During my career, I developed a keen sense of observation. In any public setting, I am always on alert, especially in a crowded area, such as a restaurant lobby. I would designate this type of awareness as an occupational trait or instinct that becomes second nature. My train of thought is always to be on alert, no matter what, and to expect the unexpected.

As I look back on that cold December morning in the restaurant lobby with my family, the sounds of pots and pans clanged, the occasional conversation between customers, and certain dialogs with the cashier as people paid before leaving could be heard throughout the restaurant. But in the midst of all the commotion, I noticed how most individuals were oblivious to everything going on in their presence. Instead, they were glued to their smartphones. These same people even ignored their children and other people in their group. I found myself wondering: *have we become so engrossed in technology that meaningful conversation with another human being is lost? Or is it that everyone forgot to brush their teeth and don't want to offend someone with their unfavorable bad breath?* Perhaps it could be a little bit of both...

Not to date myself, but I remember my first Motorola "flip phone" in the mid-'90s. The cellular phone was without a doubt a big step up from the pager that clipped on my front pocket. What a great tool! A practical, handy instrument. However, the tools have instead made life difficult.

Fast forward a few years into the future, the breakthrough technology of the Blackberry hit the market, changing the cellular phone industry forever. The next level of communication emerged with SMS texting, email, and Internet access all in one device. Stress levels in society increased yet again. But IT advancements did not stop there; the new smartphone hit the shelf. We are on the verge of new technology we never even fathomed in the past. Social media turned the restaurant lobby into a group of zombies, stuck every second to the alerts on their smartphones.

My story brings us to the point of this book. We cannot escape the unfolding world of technology and communication. It is a beneficial tool to conduct business, keep in touch with family and friends, and keep us entertained. At the end of the day, smartphones are an unbelievable means to extract information about anybody or anything. But we must draw a line in the sand to protect our identity. One must ask oneself: are we limiting the way our brains develop? Connection dependence is quickly becoming a serious issue with the advent of technology.

The constant connection has created an undetachable bond with technology, perhaps even an addiction. Is the relentless need to check our emails, texts, or social media sites causing more stress than necessary? Anyone who chooses to use the advances in technology must consider their digital security an essential part of being a responsible user. For example: if you are connected with the location settings turned on, your privacy has vanished into thin air.

In the world of private security, we build multiple levels of protection: advance teams, interior security, exterior security, perimeter security, internal CCTV, physical barriers, etc. Users are the main control tower who must determine the protection layers. The key is learning how to enact these protection protocols.

Developing the Unhackable mindset is about protecting you, your family, your business, and your assets from constantly evolving technologies such as smartphones, cloud technologies, and much more. At face value, one could say, "Oh, it's just a cell phone." But on the other hand, it is a tool that can be used for very negative things, such as cyberbullying, identity theft, or even, in extreme cases, the manipulation of the democratic process.

Society as a whole is changing before our eyes due to dependency on smart technology. Society no longer thinks independently, without the use of a machine to direct a decision. A classic example is the use of maps. I am a native-born Bostonian and know the city quite well; however, in most cases, the first thing I do is connect to my favorite map application. The technology gives me the quickest route possible and eliminates any process of thought because we have been hard-wired to depend on the tool. It is a double-edged sword. I reach my destination on time, but my brain is turned into a technologically dependent organ. The conscious thought of figuring out the route on my own does not exist anymore.

George Mansour takes the reader on a new journey unlike any teaching about technology. The concept instructs the user on how to develop an Unhackable mindset in a society filled with data breaches. Buckle up and prepare yourself for an eye-opening experience that no one has dared to uncover in history!

—John C. Guerini

Security Operative & Consultant

Retired Massachusetts Probation Officer

PREFACE

I'm fortunate because my work as a cybersecurity professional has allowed me to help people who have been hacked or have had their digital data compromised due to security flaws. For over twenty years, this has been my passion. While I have loved educating and empowering clients to better manage their online digital lives, it has been troubling to see how many people are struggling with security risks in their personal and business lives because they don't understand the foundations of digital technology. It appears users are embracing technology without learning to recognize the dangers it poses.

In the initial years of starting my business, I noticed a lack of concern for online security. The only consideration was the cost for someone else to fix the problem. Users were not interested in learning how to prevent the situation.

As the issues progressed, I became frustrated with repairing the same problem on a regular basis. The unyielding cycle expanded to a global situation where breaches spread beyond individual users or independent companies to governmental agencies. No one was slowing down to ask, "What is my exposure? What is happening with my data?"

The effort involves time, something society has not been

willing to invest. Corporations and businesses alike typically apply band-aids to solve IT-related problems. Meanwhile, the loss of private data continues to be a concern. Sadly, online security has been relegated to ventures in sales and installation as an attempt to "guarantee" security. Infrastructure is not the primary challenge. Educating users on the risks associated with hackers and scams should be the primary concern.

Avoiding technology completely is impossible. We must find a balance between ourselves and our technology. Only then can we learn to control the effects it has on our personal lives.

As a result of my work in cybersecurity, I began to question the future and wanted to create something better for my children. Since our offspring will be the ones navigating technology, it must be the parent's responsibility to educate them properly on navigating the Internet effectively. It's important they learn how to enhance, not diminish, their lives with technology.

No matter how much my clients and I discussed security protocols, it was going in one ear and out the other, unless it had to do with ROI or an emergency. Until it affected them. It became clear the problem was more severe than I anticipated. As a hired professional, it was my responsibility to protect their data privacy, hence the reason for creating the Unhackable mindset. If the future of cybersecurity was going to change, then people needed to be made aware of protocols for protecting themselves and their businesses.

In the first few years of entering the cybersecurity world, I quickly saw how my clients only focused on fixing the problems, not preventing them. Later, it became apparent the businesses, clients, and customers were lacking in these main areas: people, resources, and time. Unlike driver's education before getting a license to drive a car, there are no required classes to help users safely navigate the Internet. Yet we complain about foul play before asking ourselves, *why did the breach happen in the first place?* The problem is rampant among Internet users all over the world.

It's time to educate people on what is happening and how to stop the problem.

The plan started with my clients, and as their education grew, they learned how to protect themselves and be more mindful and present. I started to recognize trends in their mindsets around the technology that kept them vulnerable to cyberattacks. As they continued to make the same mistakes and patterns repeatedly, I discovered a few simple strategies to share with my clients that prevented them from being hacked. Since this is an issue that affects everyone on the planet, I knew the lessons couldn't stop with just my clients. My knowledge and message had to be shared with the world.

Users are stuck in a dangerous cycle with no way to regain control over their lives. Every person has negative experiences related to cybersecurity. Often, the situation may feel overwhelming without any way to take back control of your digital identities. I am here to tell you that is not true. It's time for humans to balance their relationship with machines and neutralize threats at the same time.

As with many of the people I have helped over the years, you may be concerned with the growing use of technology in society. How can you interact safely with these technological innovations, while ensuring your protection at the same time?

INTRODUCTION

"Insanity is doing the same thing over and over and expecting different results."
—Albert Einstein

How do we maintain our privacy and security when everything online is being exposed and logged? How do we secure our connections, permissions, and access to sensitive data in an unstable environment? These questions are concerns for everyone around the world. Our data information is continuously being tracked and stored.

If you are like the many others I help, you may also be alarmed about the growing web of technology surrounding us. Although, the real question is, how do we best protect ourselves against these technological innovations?

We must first remember that technology is only a tool. As with any instrument, the operator must take great caution when using the device for privacy and security reasons. We are in danger of allowing the tool an unnecessary amount of control over our lives. Therefore, we need to re-balance our relationship with technology. Education of the digital landscape and the human experience must be administered.

On the Web, but Not of the Web

The Internet is constantly being tweaked and altered to monitor online behavior. These modifications follow surfing habits that model specific tastes while gaining insight into opinions. Therefore, it only shows what you want to see.

Have you ever stopped to consider how much consumerism and convenience are really costing us? We have become intertwined in the web of cyberspace; it is slowly reprogramming our thought process to an involuntary, unconscious state of mind. Is this why we created technology?

It is time to liberate ourselves. The Unhackable paradigm shift is a process that enables a balance between new technology and the existing human condition. We can be on the web but not of the web. *Unhackable* adds a new dimension to the user's thinking and brings forth a new school of thought that unravels the complicated nature of the Internet. The digital transformation is not just about technology, but about how we create a viable partnership with technology.

The rapid development forced society to adapt in an uncontrolled fashion, leaving users unequipped to process the vast amounts of information-sharing over the Internet.

The development of technology cannot be stopped, but users can change the way they deal with advancing innovations. Are you ready to take the challenge? Privacy and security are the two main goals of protecting our sensitive data.

Freedom and the Web

As the Internet celebrates its 30th anniversary, we must ask ourselves as responsible users, are we doing everything in our power to understand this new environment? The innovation has granted incredible access to information that previous generations could have only imagined. A comparable analogy would be an ocean. Could you traverse the sea in a sailboat without proper navigation? *Unhackable* provides a map that allows users to securely protect their sensitive data.

A Formula for Freedom

Unhackable was created to build a bridge across the treacherous ocean of the Internet. It provides a methodology that frees its users from the stickiness of the World Wide Web. We can once again gain a sense of autonomy with advancing technology in society.

The concept is a timeless solution to carry us through the landscapes of our digital age. Harmony will be restored with our connections to media influences.

The Internet impacts every area of daily life. The result brings forth the knowledge to navigate technology. Connecting + Sharing + Growing = Freedom.

You are the source of power, truth, and a balanced life. To find stability in this electronic age, we must define our equilibrium, social media symmetry, and interconnected parity. Our trail of sensitive data can be adjusted either intentionally or unintentionally while using the Internet.

It is inevitable that everyone at one point will be affected by cybersecurity issues in some fashion. None of us, myself included, are completely safe from hackers. With big companies like Equifax being hacked, our personal information is always vulnerable. We can't control all of the personal information that is out there with businesses, but with Unhackable mindsets, individually and collectively, we will build those walls of privacy and security. We can start by being proactive and thinking carefully before sending or receiving online data. We need solutions to restore, reconstruct, and rebuild cybersecurity within our own lives. A "do-it-yourself" approach to breaking free from our Internet content. In other words, a natural treatment for our technology problem.

When someone repeats the same process continuously, the

actions become instinctive habits. Therefore, we must regain control of our digital connection to understand the underlying problems. The solution lies within every person using a digital device.

We must become the practitioner of our problems and begin mending each facet of the interconnected life. It is the equivalent to becoming your own virtual assistant. Only then can we create a happy and healthy existence with technology.

Your Digital Footprint

The whole business Internet model is broken; our free information-sharing highway has been hacked.

The original Internet architects, engineers, and planners created a platform for people to share knowledge around the world. It was supposed to be an environment that could connect people from different cultures, races, business associates and such. Instead, digital information control is seeking to destroy this delicate multi-layered database.

It does not matter if the technology is automatic or manually driven, you must know how the system is being manipulated to protect it from hackers. Therefore, the Unhackable mindset must be enacted. If the user does not understand the core processes of their system, then how can it be defended?

Users have a unique digital footprint that can be adjusted by using privacy and security settings correctly. *Unhackable* teaches users how to bring balance through intelligent actions while using Internet connections. It is the one controllable element that will put us back on course.

Unhackable presents the true genuine cybersecurity solution that works on the interaction between the user and technology.

PART I

AWAKENING

Reevaluate Your Relationship with Technology

1

AWARENESS IS EVERYTHING

"When you change the way you look at things, the things you look at change."
—Wayne Dyer

The most important step to becoming Unhackable is an awareness of your surroundings. But where does awareness come from? It comes from knowing that nothing is secure; our lives have become too complicated. We are influenced by our families, cultures, and society. However, to truly understand our place in society, we must acknowledge where we are, what we're doing, and where we should go. We must alter our perception of technology and the security challenges it places upon our daily lives. In doing so, it will question the relationship we carry with those electronic devices and how they negatively impact our state of mind.

The everyday user has no control over the Internet, but we can regulate the amount of accessible information. Since the invention of the Internet, people went from storing private data offline in a secure setting to uploading that material in an unsecured environment. In doing so, we have exposed ourselves to the world during a time of cyber warfare with no defense plan. It has become a professional hacker's heaven.

Nine out of ten people have absolutely no idea how exposed their data is to violators. I cringe whenever someone tells me they have never been hacked. Reality would tell another story; more than likely, good hackers get in and take what they want without ever being caught. Society loses billions of dollars every year due to data breaches.

Until today, nothing has changed regarding our core problems. We have just covered the issues with newer technology and introduced a more globalized problem. We lacked focus on discipline and consistency because we had no extra time, yet new technology was supposed to give us that and more, by making our life easy, fast, and interconnected anytime, anywhere, and anyplace.

- Did it do that for you today?
- Do you have extra time?

The technology that is supposed to make our lives safer and more freedom-based has created a whole new world of issues. You should be asking yourself: what is the purpose of all this digital innovation? The technology itself has become more dangerous than any weapon we have ever created.

We must wake up! When the Internet became widely used, it was referred to as a revolution of the digital age, connecting society worldwide. It was a time to rejoice over the ability to interact seamlessly without boundaries. Underdeveloped countries could join the 21st century. Humanity could grow through invisible borders. But the invention brought about an unknown risk. A danger that has cost much more than money, it's diminishing the conscious thought among humans. The time for transformation has become a necessity in modern times. Our awareness has faltered; we are being nurtured to depend solely on technology instead of human consciousness.

The process to rebuild our awareness is education. We are taking measures to secure our lives offline, so why would connecting online be any different? It's imperative we know what's happening with our electronic connections at all times to avoid any dark alleys in cyberspace. When traveling in unfamiliar territory, it is important to follow the road signs for directions.

Technology in Our Everyday Lives

Most of us go online and choose to do things over the Internet because it's quick and easy; we choose convenience over security. We all love technology, but we must also control digital innovations, we cannot allow them to control us.

Technological advancements continue to make everything easier by making Internet access a constant connection. In the past, we had to sit behind a desk to use a computer, but with the creation of portable devices, connection follows you everywhere.

Advances in technology are happening so quickly, nobody is slowing down to think about how technology is impacting our lives. The *Next Web* article "The Incredible Growth of the Internet Over the Past Five Years—Explained in Detail" clearly identifies just how quickly the Internet and social media sites are taking over the world:

Internet users have grown by 82%, or almost 1.7 billion people, since January 2012. That translates to almost 1 million new users each day, or more than 10 *new users every second*;

More than 1.3 billion people started using social media – that's a rise of 88% in just five years, and equates to more than *8 new users every second*;

The number of mobile connections in use grew by a whopping 2.2 billion, meaning that operators activated a net average of almost *14 new subscriptions every second* to deliver growth of 37%;

We've only been publishing mobile social media user numbers since January 2015, but users have grown by more than 50% in those two years alone. More than 864 million people have started using social platforms via a mobile device in the past 24 months, at a rate of almost *14 new users every second.*[1]

It is estimated that four billion people and over forty billion devices will be connected to the Internet.[2] That is huge growth! Everyone is embracing technology, but very few truly understand the danger it presents.

Keeping Up with the Dangers of Technology

"Computer security, cybersecurity or information technology security (IT security) is the protection of computer systems from theft or damage to their hardware, software or electronic data, as well as from disruption or misdirection of the services they provide."[3] Although digital technology use continues to increase, many users do not understand cybersecurity or why it is important to our safety. The field is expanding rapidly, due to our reliance on computing systems like IoT (Internet of Things), which is adding a new set of problems for every area of life. Society is being pushed to merge with smart technology without providing the education to consolidate or secure our data, making it defenseless before switching to the latest phone, gadget, or updates. These inventions also opened the doors for professional hackers to steal precious data from unsuspecting users. The development of a proper security system takes time and discipline. Many of the technological advancements were created to reduce conflicts, improve health, or procure financial security for many people. Our impulse to feed the instant gratification eliminates any chance of truly settling, learning, and enjoying the advancements.

Recently, a client contacted me and asked for some help reclaiming data from her photo shoot. Sally initially called her phone carrier and asked for assistance, but they were unable to recover the photos. She was informed, "You need to go see an expert in data recovery." Sally contacted me through a friend, as she was in a state of panic. It was determined that her data was completely disorganized. The information was scattered across multiple devices stored online, but was not unrecoverable. The primary concern should be to keep your data in one central location. Once I was able to organize her material, the images were recovered, thereby proving the importance of storing data in a structured location.

- How can you better manage your digital life?

In times of trouble, users need to depend on themselves for the solution. We have become so accustomed to technology that the rapid changes are overlooked. The constant struggle to keep up has been nurtured into daily behavior. To establish a management process that safeguards digital life, you must depend on the human connection. We are blinded to the issues facing society. Individuals forget what it means to think before they act.

- Has technology made your life easier or harder?

These technological advancements are supposed to help us organize our lives, but often these tools end up creating new conflicts. Smartphones were supposed to help us multitask and get more done faster and easier; instead users are distracted with multitasking on their devices, stopping them from completing one simple task before moving to the next. We are so accustomed to technology, but it is all changing so quickly that by the time we feel comfortable and have mastered the problem, we are at the bottom level, learning all over again.

Data supports my theory: a recent survey completed by The Pew Center for Internet, Science, and Technology clearly demonstrates how unaware the majority of people are about Internet safety. *The Security Ledger* analyzed the data from the survey and the results are very concerning:

> U.S. adults may be able to identify a strong password when they see one, but on many questions of how to identify and protect themselves from online threats, they are worryingly ignorant.
>
> Less than half of Americans knew what **ransomware** was or understood that their email messages and wireless traffic are not **encrypted** by default. Just 16 percent correctly identified a description of a "**botnet**." And, in an age of widespread password theft and account takeovers, just 10 percent of adults could identify an example of **multi-factor authentication**. The results have implications for employers and public policy experts alike, as sophisticated cyber-attacks often rely on so-called "**social engineering**" attacks on individuals, especially in their earliest stages.[4]

Protection from hackers is impossible if you are not informed. Anyone can help secure themselves if they understand the basic terminology and are aware of how it can be used to either hurt or protect them.

Dependency—Technology controls us because of our dependency.

- If your phone is taken away, are you lost without it?
- Do you know how to use a paper map?

To develop an Unhackable mindset, the awareness of our dependency is imperative.

As technology use increases, we will become reliant if

something does not change. Innovation was meant to improve our lives, but instead, it has created Internet dependency that interferes with many people's lives. Even if you are not clinically diagnosed with Internet Addiction Disorder, chances are, you still use social media more often than you should. The constant connection makes users vulnerable to hackers or businesses who seek to take data for profit. The addictive behavior isn't entirely your fault, and it isn't an accident. Chapter 3—Modern Data Gathering will explain how and why we are so drawn to social media.

Now that you have a better awareness of our role in technology and how it plays into everyday life, it's time to learn how society became so dependent on technology.

OUR CHANGING RELATIONSHIP WITH TECHNOLOGY

"It has become appallingly obvious that technology has exceeded our humanity."
—Albert Einstein

Our mission is urgent since the control of privacy and security has been lost due to nation-state-sponsored hacking, criminal behavior, ad-based revenue models, and misinformation. Technology interaction affects everyone, as we are interwoven over the Internet as one, but individually unique. The Unhackable mindset creates the conceptual model to enact a multi-layer digital union between technology and human interaction. The insecurities and impurities that threaten us online with every new connection must be controlled through a collaborative approach that will suppress the threats to our sensitive data. It's time to reclaim our digital lives and digital freedom.

The major influencers within the IT industry took the world by storm with their groundbreaking innovations. Followers took advantage of their innovations without thinking critically about the personal and social consequences of using the new technology almost incessantly.

The Beginning with Bill Gates

In 1974, Bill Gates read about a personal computer kit, Altair 8800, and felt it was a necessary item for the future of humanity. It was all about timing. Gates and his partner Paul Allen knew they could improve the personal computer by using the proper processing language.

"Allen spoke to Ed Roberts, president of Altair manufacturer MITS (Micro Instrumentation and Telemetry Systems), and sold him on the idea. Gates and Allen worked night and day to complete the first microcomputer Basic."[1] Gates anticipated the change that was coming and dropped out of Harvard to focus on the future of personal computing.

In 1980, IBM contacted Microsoft about an operating system for their personal computer. Gates took a big risk because he didn't have the software at that time. Instead, he quickly bought another company's software and presented it to IBM. "For this deal, Microsoft bought a system called 86-DOS from a company called Seattle Computer Products and, after adapting it for the PC, delivered it to IBM as 'PC DOS' in exchange for a one-time fee of $50,000."[2] In one transaction, the world of technology forever changed.

Bill Gates saw the future in the first personal computer. However, could he have seen the concerns with the intelligence and information that would come from new innovation? With the advancement of new technology, it contributed to the privacy concerns that we have today. However, as technology became user-friendly, it also became a powerful tool for data gathering.

Microsoft became a giant due to the mouse and GUI (graphical user interface). Early computers used command prompts and only a select few knew the proper language.

Microsoft brought technology to the masses, but nobody took the necessary classes to learn the language.

Humans, as a society, didn't focus on education or training, we just leaped without thinking about the risk. Adults must obtain a driver's license to have a car, doctors need a degree to practice, why should users not need a certificate to operate a digital device?

As an innovator, Gates is constantly thinking of future advances in technology. He continues to encourage society to use more advanced devices. Currently, Gates is working with Belmont Partners to create a "smart city" in Arizona. Business Insider's "Bill Gates Just Bought 25,000 Acres in Arizona to Build a New 'Smart City'"[3] shares, "Belmont Partners expects its development to feature all the trappings of a futuristic city: high-speed Internet embedded in the built environment, accommodations for self-driving cars (such as traffic lights that communicate with one another to minimize congestion), and smarter manufacturing technology."

However, the impact of all the technology on the future residents of Belmont is unknown. Brooks Rainwater, the director of the City Solutions and Applied Research Center at the National League of Cities said, "When details are released, he hopes technology will serve as the 'backbone' of the city, 'not the purpose of its existence.'" Are we focusing on the tech itself, rather than on the quality of the lives for those using the technology?

Our Obsession with Apple

Apple has been incredibly successful in the tech industry. Apple has had an almost fad-like following as fans fervently anticipate new products, such as iPods, iPads, and iPhones. The *LiveScience* article, "Apple Obsession: The Science of iPad Fanaticism" explores just how fanatical some followers became:

Some fans have retold and romanticized Apple's history, giving a "legend" to the company and its founders. In 2005, Belk and his colleagues found evidence for several myths within the Apple community, including a "creation myth" involving the creation of one of the first computers in Apple founder and CEO Steve Jobs' garage.[4]

Steve Jobs made a major impact on our lives with technology. Historically, Apple has been ahead of everyone else with tech development and trends; however, with iPad sales declining, it seems like Apple may have to rethink its plan and adjust to the ever-changing environment. One of the reasons for this decline is that people may be starting to think more critically about whether they actually need a certain type of technology. We don't blindly buy everything Apple and other technology companies produce. In "The World of Technology Is Changing and the iPad Is Getting Caught in the Middle," Matt Weinberger shares:

> Apple CEO Tim Cook talks the iPad up as a product with a bright future ahead. But with the decline in sales so unmistakable, that future is getting called into question...
>
> And so, the question becomes less, 'Do I need a tablet?' or 'Will this tablet replace my laptop?' and more like, 'does this device do what I need it to do?' It's a healthy reminder that technology is supposed to serve you and your needs, not the other way around.[5]

Not only are sales down, but Apple is on shaky ground; they are realizing that their system is not as great as they thought. An increase in security attacks against Apple and Macintosh has caused concerns with users. "Eugene Kaspersky, the CEO of security firm Kaspersky Lab, says Apple is headed for a rough patch... 'Cyber criminals have now recognized that Mac is an interesting area. Now we have more malware, it's not just

Flashback or Flashfake,' Kaspersky told CBR. 'Welcome to Microsoft's world, Mac. It's full of malware.'"[6]

Unfortunately, we are moving in a direction that focuses more on automation and profit over human connections and fair business practices. Influencers are reaping the benefits while we're in danger, and that's wrong! Steve Jobs contributed greatly to technology, but he accomplished a lot through smart marketing practices.

"He had a unique way of crafting his own reality, a 'distortion field' he used to persuade people that his personal beliefs were actually facts, which is how he pushed his companies forward. He also used a blend of manipulative tactics to ensure his victories, particularly in boardroom meetings with some of the most powerful company executives in the world."[7]

Pushing New Versions

When a program tells users an update is available, we need to ask why the update is necessary. Is it due to a security problem or just an enhancement update?

These same companies keep moving forward, releasing new versions, and producing more advanced innovations before they have perfected the current version. Instead of facing the current problem and security concerns, we are being blinded by the newer version, along with new issues to cover and patch the previous set of problems and bugs. We should be trying to work in a collaborative effort to create a more stabilized environment; stop adding all the innovation and stick to what is working. Users should not update in a real-time environment online. Instead, we should have a testing environment where updates can be applied and tested before it's implemented in real time. Otherwise, sensitive data becomes the guinea pig.

Apple even admitted that "it has been secretly stifling the performance of older iPhones. Critics have accused the company in the past, based on anecdotal evidence, of purposely slowing phones to compel users to upgrade to the latest model...The fact is Apple has an incentive to push users to upgrade; it makes money selling new devices, after all."[8] Apple greatly upset many of its most devoted followers when the company admitted to slowing down the batteries of older iPhones.

Business Insider's article "Apple Confirmed a Longtime Conspiracy Theory - and Gave Regular Customers a Big Reason to Distrust It" also reveals another tactic Apple has used to force users to upgrade:

> The company has a history of artificially making older devices look inferior to new ones. The iPhone 4, for example, was perfectly capable of running Siri, but Apple reserved that feature for the model that replaced it, the iPhone 4s. Likewise, the camera in the iPhone 3G was capable of shooting video, but Apple didn't turn that feature on and instead made video recording the signature capability of its next device, the iPhone 3GS. Rather than having the choice to upgrade, many felt that they were being forced to upgrade through manipulation.[9]

So, Apple gets to run our lives the way Steve Jobs ran his company. It is all about strategy, and we're held hostage; we have no choices anymore. By the time we're comfortable, we have to change again and adjust because of the company's strategy. The innovation is blinding the consumer to a point where they just use it without thinking critically about how it is impacting their lives. Companies utilize algorithms like artificial intelligence in combination with built-in stored databases and user activity. This has made everything smart, and now companies know more than the user knows himself.

Everything is connected and interrelated. It's a domino effect. A new approach needs to be initiated; stop this constant running and never finding a comfortable place to function. Users are allowing the situation to continue due to a lack of knowledge and understanding. We must stop letting the advancement of technology push us around.

Influencers have a strategic plan for us. Why do they get to decide the plan?

The major companies control the market and the advancement of innovation. However, we have great ideas and are all leaders. We are all powerful. We need to neutralize all controls with our online identities. Why do we have to become followers?

Unhackable was designed to adapt with the advancement of innovation, allowing the user more freedom to bridge the gap between technology and human interaction. Thus, becoming a leader in control and empowering those around us.

3

MODERN DATA GATHERING

"The true axis of evil in America is the brilliance of our marketing combined with the stupidity of our people."
- Bill Maher

Are you able to surf most websites if the cookies are not enabled? The answer is no. Society has generated erroneous assumptions about technology, innovation, and markets built on data promotion, sales funnels, and the art of web-surfing. The Internet is constantly being tweaked to monitor online behavior and follow user patterns, tastes, and opinions. Internet algorithms show only what an individual's digital usage concludes; displaying the results based on our regular search history.

Information is being sold for the benefit of the corporations. We are being reduced to specific algorithms in an effort to influence our decision making through calculations, data processing, and automated reasoning tasks, allowing for the subliminal manipulation of our short-term and long-term consciousness. This security threat is a concern to many people around the globe.

In "Cybersecurity Isn't Just an IT Problem—It's Also a Marketing Problem," Cybervista CMO Lynne Koreman shares:

> Marketers ask the right questions about privacy and protection, but sadly they also ask, "How fast can we go with implementation?"
>
> Driving revenue and consumer interest is always top of mind for marketers, but considerations around protecting customer data must come first...It should never be "us vs. them."[1]

The data sets and tools that marketers use are prime targets for hackers, increasing the threat against our privacy and security. Marketing teams must find alternate methods to avoid the manipulation of user's predictive analysis. Marketers know who the potential customer might be, then gather and analyze their usage habits.

Businesses cannot connect with target customers unless they are familiar with their spending behaviors. When the analyzer casts a wider net, they can better resonate with their target markets. Once the company knows the consumer, their mined data is used to shape branding methods and increase profits. In this situation, advertising expenses are far cheaper than most methods of promoting. Facebook, social media, and Google's targeted ads prove that a little can go a long way with this type of system.

Companies know how to leverage your data to successfully improve user interactions. This is the reason businesses need you to always increase, click, and play, so they can adjust, measure, and prune feedback. The root of all successful digital companies is customer data. The analysis allows the company to sell or serve their clients more effectively. The data is based on our personal habits.

Instead of us surfing to pick what we want, we're guided to

see exactly what our data says we love, so that companies may target and sell to us more and more. We don't see that we're losing control to satisfy their needs. When shopping in a mall, you are able to see options that lead to the choices for a particular need. With online predictive coding, the choices are determined by your actions, thus narrowing search results.

When will it stop? It starts and ends with you—the user. We must learn to make conscious decisions with our best interests in mind. Unfortunately, as businesses increase sales, do they realize they are sacrificing our mindsets, the fabric of our identities and our fundamental rights? We are sacrificed for the numbers and their projections. In the digital age, the user's every move is tracked, so companies know what and how to send out their marketing.

Data Mining

Target knows so much information about its customers that the company can make predictions about when customers will get married, have kids, and more. The marketing information sent out is based on those predictions. Use of their complex pregnancy-prediction model has led to Target predicting a woman's pregnancy before she'd even announced it. One incident was when a father learned of his daughter's pregnancy through Target's tracking. The *New York Times*' article "How Companies Learn Your Secrets" shares this anecdote:

> A man walked into a Target outside Minneapolis and demanded to see the manager. He was clutching coupons that had been sent to his daughter, and he was angry, according to an employee who participated in the conversation.
> "My daughter got this in the mail!" he said. "She's still in high school, and you're sending her coupons for baby clothes and cribs? Are you trying to encourage her to get pregnant?"
> The manager didn't have any idea what the man was talking

about. He looked at the mailer. Sure enough, it was addressed to the man's daughter and contained advertisements for maternity clothing, nursery furniture and pictures of smiling infants. The manager apologized and then called a few days later to apologize again.

On the phone, though, the father was somewhat abashed. "I had a talk with my daughter," he said. "It turns out there's been some activities in my house I haven't been completely aware of. She's due in August. I owe you an apology."[2]

Target's statistician Andrew Pole candidly discussed what information Target gathers:

Pole outlined exactly which information Target collects for each guest to associate with his or her "Guest ID" (the number that Target uses to track its shoppers). It starts with name, address and tender (the credit card or debit card you use) and expands from there to a history of your store purchases, online purchases, mobile phone ID, actions taken in response to Target emails, and Internet browsing activity if you click on a link in one of those emails. Location data is interesting to the company; if you live right near a competitor, they'll try to steer you toward shopping at Target.com.[3]

Target is far from the only company that mines data.

Companies' ability to track people online has significantly outpaced the cultural norms and expectations of privacy. It is not because online companies are worse than their offline counterparts, but rather because what they can do is *so, so different*. We don't have a language for talking about how these companies function or how our society should deal with them.[4]

Current technology and outdated privacy laws allow tech-savvy companies to gather so much data that they might know more about your life than you do!

Users are moving too fast as businesses sacrifice our safety in their quest for profits. Sales are made at the cost of consumers' privacy. Companies are focused on getting their products online and selling to meet the predictions. Customers are just part of the financial equation. The human connection is irrelevant— along with security of the data. Guess who gets caught in the crossfire? We do.

Inbound marketing is putting users in danger: "If you think companies and individuals have your data under lock and key, then think again. With hackers becoming more sophisticated and rarely ever getting caught, we'll likely see more and more breaches in the future."[5]

All of your data is out there, and because of that, the hackers are flourishing in this market.

"Hackers responsible for data breaches at companies often put the information they have stolen on the dark web for others to buy and make use of for financial gain."[6]

Would you leave your house with doors and windows wide open, so anyone can come in and out? Do you want someone snooping in your closets? Do you want that with your technology? Your data, your email, all your notes, everything you've utilized online is being monitored. It's all being controlled. Would you rather not have control over your information and keep it under lock and key?

Most of our security problems come from exposed data that initiates a breach. "A data breach is an incident in which sensitive, protected or confidential data has potentially been viewed, stolen or used by an individual unauthorized to do so. Data breaches may involve personal health information (PHI), personally identifiable information (PII), trade secrets or intellectual property."[7]

Have you ever stopped to consider how much our end goal of consumerism and convenience is really costing us? Is it worth our lives? We are intertwined in a technological web that is slowly reprogramming the way our minds think and react. Technology is taking our autonomous human minds and changing them into instinct-based animal minds. Is this why we created technology?

It all starts and ends here with the Unhackable mindset.

Psychology of Marketing

Patterns. Pictures. Patterns. Pictures. When we see something enough, we start to believe it. Which is why marketing companies are looking to see how they can push their products to potential buyers. They know how our brains function. In *Rhetoric*, Aristotle's seven causes of human action reveal that we are often driven by compulsions, habits, and desires. Marketers are using that information against us. "Consumers might change the way they like to do things, make purchases, gather information, and spend their time, but their basic psychological needs and philosophical causes of action are the constants that marketers can always count on."[8]

Look at how much businesses spend on commercials. McDonald's spends the most on ads with the intention of creating an impulsive action. "Globally-known quick service restaurant (QSR) chain McDonald's spent approximately 447.3 million U.S. dollars on advertising worldwide in 2019."[9] We are being nurtured to spend money frivolously, without thought as to the end result.

Modern marketing relies so heavily on potential customer data that companies are willing to buy data if they cannot collect it themselves from users. "Data brokers are entities that collect information about consumers, and then sell that data (or analytic scores, or classifications made based on that data) to other data brokers, companies, and/or individuals. These data brokers do not have a direct relationship with the people they're collecting data on, so most people aren't even aware that the data is even being collected."[10]

Any bad actor can capitalize on the use of your data, thanks to the creation of inbound digital marketing. It's not about what's best for the user anymore, but we can change the future by getting back in the driver's seat and re-taking control.

Feeding Your Compulsion

Our compulsion is not an accident; it was created by design with apps and social media fashioned to increase financial gain. Be cautious. If the product is free, users may become the service. Using the proper configurations and settings can protect your safety. Technology generates an environment to predict addictive behavior. Increased online usage gathers more data for businesses to force specific advertisements to their potential customers.

Center for Humane Technology, a non-profit organization focused on creating harmony between technology and humanity, points out that:

> Facebook, Twitter, Instagram, Google have produced amazing products that have benefited the world enormously. But these companies are also caught in a **zero-sum race for our finite attention**, which they need to make money. Constantly forced to outperform their competitors, they must use increasingly persuasive techniques to keep us glued. They point AI-driven news feeds, content, and notifications at our minds, continually learning how to hook us more deeply—from our own behavior.[11]

The founder of Center for Humane Technology, Tristan Harris, has a unique insight into how social media manipulates our minds to meet its goals. Harris spent three years as a Google Design Ethicist, focusing on the ethics of technology. Through his work and research, he has recognized many of the strategies that sites use to keep us hooked. Product designers "play your psychological vulnerabilities (consciously and unconsciously) against you in the race to grab your attention."[12] They control what we see, provide incentives to keep us engaged, encourage our concerns about social approval and missing out, and constantly interrupt us with notifications to draw us back in.

The joy that technology brings you has been carefully crafted in a way that keeps you hooked. Apps become addictive by giving users incentives at specific times that make them want to spend more time using the app. Dopamine Labs has a "platform that uses AI and Neuroscience to personalize moments of joy in your app. It adapts the rhythm and timing of ✳s to surprise and hook each user. They'll stay longer and engage more. Up to 167% more."[13] The company created this technology to help users thrive by working with apps that encourage overall health and wellness. "The relationship between humans and our machines must be one of mutual thriving and improvement. The little magical slabs of glass that live in our pockets and proliferate across the globe are tools for human thriving."[14] However, the concern remains because the user's freedom to make independent choices is being manipulated by algorithms.

The expert marketers are outsmarting everyone. At the end of the day, as the desire for instant gratification increases, so does the demand for more data collection. Center for Humane Technology presents a few strategies for disconnecting and taking back control. Two steps to take are to charge your phone outside of the bedroom and only allow notifications from people, "Most notifications are from machines, not actual people. They keep our phones vibrating to lure us back into apps we don't really need to be in."[15] Limit the number of apps you use and create boundaries for yourself that will help you resist the allure of likes, tweets, and snaps.

Develop Your Unhackable Mindset!

Are you ready to take the first steps toward developing an Unhackable mindset? If not, keep reading; otherwise, go to your *Unhackable Workbook* and complete the exercises and questions for Part I. Within the workbook, complete a questionnaire about Internet habits, take an inventory of all the technology you use at home and work on a daily basis, and review the websites and apps that keep drawing you in.

To access your complimentary *Unhackable Workbook*, go to www.GeorgeMansour.com/workbook

Part II of *Unhackable* teaches users about the current dangers of technology. It shows the negative impact these connections have on our lives and the dangers of digital advancement.

PART II

DANGERS OF TECHNOLOGY

To You, Your Family, Your Business, and Society

4

CYBERSECURITY ISSUES

"A hero is someone who understands the responsibility that comes with his freedom."
—Bob Dylan

The importance of cybersecurity does not just apply to surfing the web, users can be attacked anytime, anywhere, and anyplace a connection is established. As an example, a client, Maria, recently received an email she thought was a supplier's invoice. The email had an attachment that looked normal; it was even titled, "New Invoice 2417-16." The sender was listed as "Carrie Shilling," clientname-inv.zip.

In the email, it stated, "This email is to inform you a new invoice has been generated for our account. Please see attachment. The file is password protected for security purposes. The password is 123456. Thank you, Carrie."

At this point, Maria had no reason to question the email. She opened the attachment and put in the password. When the file was downloaded, it encrypted the hard drive, holding her system hostage. The zip files were scripted and took over the whole computer!

It is not necessary for thieves to enter a house and physically hold you hostage. In this situation, it happened right at her office desk. The ransomware infiltrated her documents, photos, and databases, along with every other important file in the system. Once she paid the money, her computer would be released.

Maria panicked, like most anyone would, which is the reason why hackers are so successful. The site pushes a countdown, with special prices to be paid by Bitcoins only if purchased in a preset time frame. For example, "Special price of $619.00 for 4 days, 23 hours, 25 minutes, and 45 seconds." After the countdown, the price would double. Often, timers are used as a pressure tactic to keep victims from thinking about an alternate solution.

The site instructed Maria on how to create a bitcoin wallet and how to convert the money. Bitcoin's price was 0.7828 equaling $619.00. Some of these ransomware programs work under the FBI radar, due to the amount requested. The hackers know these laws extensively and work under the amount that would allow you to file a case with the FBI.

In her panicked frame of mind, Maria chose to call me, and I came over immediately to fix the problem. Unfortunately, in most cases, the only repair is to wipe the system clean if you don't pay the ransom, losing all data stored on the hard drive, unless you have a full backup. Thankfully in this situation, Maria was lucky we were able to restore her data from her backup immediately. Since she had been a client for many years, she was well-trained to back up all the data on a regular basis and to verify the backups frequently.

Are you backing up today? Do you know where all your data is right now? When was the last time you verified your backup? If you were experiencing a data failure right now, would you know where to begin and how to restore your data?

Mara had fallen victim to similar scams in the past. After

some further instruction, she learned to stop and think before she acted. The situation was contained without any further problems, and Maria started to develop an Unhackable mindset. All emails are marked with a name and email address. If you don't recognize either do not open. Delete immediately!

I used this example to prove how important it is to think before you act. The purchase of products and services will not replace human intelligence; the user is alone on the front lines as they click away. You are the first and last line of defense, the beginning and end all at the same time. Ask yourself, will you pause and give yourself time in a similar situation? Without proper protocols, threats will become breaches, and these issues will create an unsecured business or personal environment, exposing all your sensitive data. Are you willing to continue to take that risk, or will you start developing that Unhackable mindset today?

Most users aren't able to escape the Internet one way or another anymore. We are connecting through so many mediums, but what matters is we are constantly sending sensitive data to the cloud, like our location history, pictures, search history, contact information, bank accounts, passwords, logins, and social security numbers. Much more is being uploaded and transmitted to companies and platforms alike. These platforms are storing and collecting so much information. Do you trust these platforms to keep your data safe? Living in these unsecured days, users will need to become proactive, defining our shared responsibility and become more aware about better transparency and accountability.

Cyberattacks happen in so many different dimensions and forms that they require a greater awareness to stop the vulnerabilities in our personal and business environments. We need to plan for a cyberattack and make sure we are putting the correct preventative measures in place to mitigate all our data risks. Attackers can come from internal or external groups. Internal groups might be disgruntled employees or thieves.

External groups can be amateur, white, black, or grey hat hackers, organized attackers who are criminals, hacktivists, nation-states, or terrorists. There can also be coordinated attacks where different groups work together creating an overlap.

Content data is "any data whether in digital, optical, or other form, including metadata, that conveys essence, substance, information, meaning, purpose, intent, or intelligence, either singularly or when in a combined form, in either its unprocessed or processed form. Content data includes any data that conveys the meaning or substance of a communication as well as data processed, stored, or transmitted by computer programs."[1]

Here are some good questions you can ask yourself before you send and transmit any content data:

1. What content data is being accumulated and stored?
2. When will my content data get cleared?
3. Where can I view my content data?
4. Who has access and permissions to my content data?
5. Who is making the decisions based on my analytical content data?
6. Why is my content data being profiled?
7. How is my content data being delivered, used, and shared?

Ransomware

Ransomware hacks are big operations that usually ask for bitcoin in their cyber demands. The FBI's 2015 Internet Crime Report defines ransomware as "a form of malware that targets both human and technical weaknesses in organizations and individual networks in an effort to deny the availability of critical data and/or systems."[2] These types of scams are happening to everyone. The DC Police and city services, such as San Francisco's light rail system, have been hacked in this fashion. The insecurities and impurities are leaving our internal systems vulnerable to all levels of hacking.

It is important that we neutralize the speed in which we utilize the new innovations to secure the safety of our sensitive data before accessing the Internet. Our current relationship with technology allows innovation to pilot us, instead of us driving technology. The time has come to govern our own data.

Recently, the entire city of Atlanta dealt with a ransomware attack, one of the biggest city network security breaches, which resulted in major disruptions, huge recovery costs, and the permanent loss of valuable information. In "Atlanta's Ransomware Attack May Cost the City $17M," Julie Spitzer shares:

> The ransomware incident knocked out services such as warrant issuances, water requests, new inmate processing, court fee payments and online bill-pay programs across multiple city departments. To unlock the city's systems and data, hackers demanded $51,000 in bitcoin, which the city refused to pay. The full extent of the damage is not yet clear, although AJC and Channel 2 Action News discovered two months ago that years of Atlanta Police footage from officers' patrol cars were lost and unrecoverable as a result of the incident.[3]

As cities increasingly rely on digital technology, they make themselves more vulnerable to attacks by hackers who easily identify security vulnerabilities and exploit them for profit. *TechNewsWorld* presents some alarming information about the ransomware trend that is only getting worse:

> Ransomware has become a lucrative pursuit for hackers, which is why it will continue to be a problem. An estimated billion dollars will be paid to digital extortionists in 2016, according to the Herjavec Group.
>
> "Hackers have every incentive in the world to continue these

attacks based on unsecure defensive systems," said Mark Dufresne, director of threat research and adversary prevention at Endgame.[4]

Businesses are also attacked often, which causes great economic damage. *TechNewsWorld* also published an article on global cybersecurity. In it, Microsoft shares that "Seventy-four percent of the world's businesses expect to be hacked every year, with the economic losses from cybercrime averaging US $3 trillion per year."[5]

Business owners are under pressure, as most are unprepared without the proper financial budgets and cost planning; therefore, they must act quickly to ensure sovereignty over their privacy and security. The lack of decisions and planning puts the business, employees, and client's information in great danger. Careful consideration must be taken to ensure and develop an Unhackable mindset individually and within every environment as a whole, so that we can all restore privacy and security back into place. No one is immune to a breach.

Professional hackers are successful, due to our lack of knowledge and understanding. The inadequacy of users taking responsibility for their connection will guarantee job security. Cyber attackers are very resourceful; they target your email. By sending a carefully created fake email from a familiar business or friend, they can fool consumers into opening the message or attachment. At that point, the computer is held ransom.

Hackers outright threaten users by using the "fear factor" and trying to reveal private information from public databases in an effort to blackmail them. The following email has made an appearance in many inboxes and many similar scams fool users daily.

From:
Sent:
To:
Subject:

Hello,

I am a spyware software developer. Your account has been hacked by me in the summer of 2018.

I understand that it is hard to believe, but here is my evidence (I sent you this email from your account).

The hacking was carried out using a hardware vulnerability through which you went online (Cisco router, vulnerability CVE-2018-0296).

I went around the security system in the router, installed an exploit there. When you went online, my exploit downloaded my malicious code (rootkit) to your device. This is driver software, I constantly updated it, so your antivirus is silent all time.

Since then I have been following you (I can connect to your device via the VNC protocol). That is, I can see absolutely everything that you do, view and download your files and any data to yourself. I also have access to the camera on your device, and I periodically take photos and videos with you.

At the moment, I have harvested a solid dirt... on you... I saved all your email and chats from your messengers. I also saved the entire history of the sites you visit.

I note that it is useless to change the passwords. My malware update passwords from your accounts every time.

I know what you like hard funs (adult sites). Oh, yes ... I'm know your secret life, which you are hiding from everyone. Oh my God, what are your like... I saw THIS ... Oh, you are a dirty naughty person ... :)

I took photos and videos of your most passionate funs with adult content, and synchronized them in real time with the image of your camera. Believe it turned out very high quality! So, to the business! I'm sure you don't want to show these files and visiting history to all your contacts.

Transfer $933 to my Bitcoin cryptocurrency wallet: 1Lmk4eUXcmtVU6YQvaPJ4yihu4fEcKtkby Just copy and paste the wallet number when transferring. If you do not know how to do this - ask Google.

My system automatically recognizes the translation. As soon as the specified amount is received, all your data will be destroyed from my server, and the rootkit will be automatically removed from your system. Do not worry, I really will delete everything, since I am 'working' with many people who have fallen into your position. You will only have to inform your provider about the vulnerabilities in the router so that other hackers will not use it.

Since opening this letter, you have 48 hours. If funds not will be received, after the specified time has elapsed, the disk of your device will be formatted, and from my server will automatically send email and SMS to all your contacts with compromising material.

I advise you to remain prudent and not engage in nonsense (all files on my server).

Good luck!

If you receive an email similar to the one above, don't panic. Please ask yourself these questions before you act to protect the safety of your content data.

- Are my actions within the law?
- Have my previous actions exposed personal data to unscrupulous individuals?
- Has my behavior led me into illegal websites or areas on the Internet?
- Have we been redirected into disreputable situations?

Users must make preparations before connecting to the Internet in order to regulate a safe environment for their sensitive data. Have you set up and configured user technology with the correct protocols and safety in place? Are you keeping applications, software, and hardware firmware up to date?

Prior to using any product services, subscriptions, IoT, and apps, users must learn how to properly configure and install before any usage. Have you applied the appropriate policies and settings to protect your privacy and security by design?

Email spam filtering is a must when it comes to regulating

unnecessary phishing mail. Once settings are adapted to your email service, it will limit the amount of junk mail, helping to protect sensitive data from hackers trying to phish and steal personal information.

If users receive an email from themselves stating their email has been compromised in any way, call the email vendor and ask if your email domain address can be blocked from the outside world. This will minimize the emails that are pretending to come from you.

As high-value targets are hacked, and our personal data like passwords are being leaked and published online, hackers are using these public databases in other types of attacks and threats to trick the user into thinking they have been compromised when they really weren't. Therefore, you need to stay afloat of all breaches that may have your accounts listed, so that you can properly secure your online identity.

On June 5, 2012, LinkedIn was hacked and over 6.5 million user accounts were frozen due to the breach. The only way to contain the situation was to lock access to the user accounts. LinkedIn officials recommended account holders change passwords on a regular basis to help protect their information.[6] You never want to use the same passwords on every site. If you are unaware of these recent breaches and you never changed your password or added any extra security layers, then you must take precautions to protect yourself immediately.

There are services and subscriptions to search the public databases for existing content about the user. You can become proactive in protecting against security threats, like the email above.

Here are some options to help secure your online accounts:

- Choose a strong password
- Use multifactor authentication
- Use a different password for each online account
- Don't share passwords with anyone

- Change passwords often

When users do not use recommended authenticated log-in suggestions, it makes them easy targets. Your decisions allow for the development of an Unhackable mindset.

In Part III, you will learn additional steps to protect yourself. Once you have gained a complete awareness and understanding of what to look for, you can start to develop that Unhackable mindset.

Cybercriminals

Dangers persist in every area of online connection, including email attachments like the one Maria opened; however, there are many ways for hackers to access data and invade your privacy. One major culprit is the use of apps, especially flashlight apps, which collect your information once downloaded. According to a *Tech Times* article, "Many of the apps have the ability to read phone status and identity, view Wi-Fi connections, modify system settings, obtain full network access, and determine your precise location via your phone's GPS, among other permissions."[7] The information can be shared with cybercriminals and puts you at great risk. Even tech experts, such as Elon Musk's former tech guru Branden Spikes, are seriously concerned over hacker's recent behavior:

> Spikes helped pioneer and build extraordinary technology, and now works as the technology evangelist at Cyberinc. But he's worried that most people don't know enough about the dangers they face online. Hacking has gone from IT whiz kids testing their prowess to a Mafia-like extortion with the potential to cause cataclysmic damage to companies and the economy.[8]

Hackers are getting through even the strongest security software. John McAfee, the founder of the McAfee computer security software, has even been hacked. His phone number was hijacked and his Twitter account used to falsely promote specific cryptocurrencies. "Cybersecurity guru John McAfee, who courted online ridicule after he was breached by a hacker seeking to promote obscure digital coins, told RT we are essentially defenseless in the face of advanced hacking techniques."[9] One of the experts in security realizes that no software can truly protect us.

Cybercriminals are always coming up with new, sneaky scams to use against us. One scam is to pretend to be tech support for your computer. They act like they are diagnosing a problem and will ask for payment to fix the problem. Alternatively, they might put ransomware on the computer. Lorrie Faith Cranor, the associate department head of Carnegie Mellon University's Department of Engineering and Public Policy, warns that "Companies like Microsoft are not actually going to call you to tell you about problems with your computer. If somebody calls you to tell you they're from Microsoft, don't believe them."[10] She recommends caution when communicating with any unsolicited tech support.

Another hidden scam has been found in popular quizzes on social media. Many quizzes on sites like Facebook seem like harmless fun, but hackers can use them to gather your personal data. Sri Sridharan, with the Florida Center for Cybersecurity, "acknowledges that it seems harmless, but you never know who is really asking you for this information."

Hackers can access data through your Facebook account and "can even trick you into downloading malware." Sridharan shared some recommendations from the Florida Center for Cybersecurity with NBC News, which includes being cautious of quizzes that want you to sign in, treating your email address like cash, not always trusting links from friends, and only participating in quizzes from reputable companies that protect data.[11]

Many innocent people have fallen for a phone scam that tries to get a recording of the caller saying "yes" to questions like "Can you hear me?" The recording is then used by scammers to get an authorization of credit card charges. They may already have your personal information due to data breaches.[12] Who would believe that answering yes to a simple question could put you in so much danger?

Cybercriminals are also attacking corporations in this new form of cyber warfare. These attacks not only hurt businesses, but they threaten our individual freedoms as even the most powerful institutions in the world are attacked.

One of the largest breaches was with Equifax in 2017. Over roughly 147.9 million Americans had their data compromised when hackers stole personal data from the credit monitoring firm. While consumers can't control the information Equifax collects, such as social security numbers, birth dates, addresses, driver's license numbers, and credit card information, there are ways that we put ourselves at risk:

> When it comes to identity theft, you may be putting yourself at risk without realizing it.
>
> "Putting too much personal information out on social media is the most egregious example" of how consumers set themselves up, according to Jeff Faulkner, acting president and CEO of the National Foundation for Credit Counseling, or NFCC.
>
> In the wake of the Equifax data breach, such personal details like where you grew up, where you vacation and who your friends are—which are all easily found on Facebook and Instagram—may be the missing link scammers need to access your accounts, he said.[13]

How can you achieve an Unhackable mindset? As we connect across the Internet in collaboration, it is our individual presence that creates the Unhackable mindset. The one technological interaction we can control, regulate, and secure within our own environment is content data. We are interwoven over the Internet as one, but individually unique. The Unhackable mindset creates the conceptual model to enact a multi-layer digital union between technology and human interaction. The insecurities and impurities that threaten us online with every new connection must be controlled through a collaborative approach that will suppress the threats to our sensitive data. It's time to reclaim our digital lives and our digital freedom today. This augmented, real-time security is the Unhackable mindset.

We are living in an unpredictable era where specific groups and professional sectors are being targeted, and personal information is exposed publicly. What was once private is now public information on the Internet. Instances such as the Equifax breach are, unfortunately, out of the user's control. Therefore, we must learn how to protect our content data. Once the pieces of the puzzle are in place, it paints a complete picture of our movements and activities pertaining to the digital life.

The Unhackable mindset must be enacted so we can break the connection and links between our personal information and content data to stop the manipulating target of our digital life.

In the age of big data, hacks are being used to target both specific groups and professional sectors. Microsoft President Brad Smith shared Microsoft's concerns about the damage being caused to corporations and governments:

Cyberattacks historically have focused on military and economic espionage, Smith noted. However, the 2014 attack on Sony was considered revenge against the company for the unflattering depiction of North Korean dictator Kim Jong Un in a film.

While cyberattacks in 2015 involved nation-states going after companies' intellectual property, attacks in 2016 targeted various Democratic party and government institutions in the U.S., threatening the democratic process itself.[14]

Our civilian identities are being leaked due to espionage, which is now trying to control our future movements and activities for nefarious purposes. In the cyber security realm, we must form an alliance with our national government to minimize the risks and threats that are working to expose our nation and individual presence. A threat against one is a threat against us all. We must realize there are hidden elements spreading misinformation all over the Internet. Cyber warfare has become a reality. The situation will not change until users are willing to create a battle plan to implement the correct modifications.

Modern innovation has brought forth devices that have created a cluster of content data fueled by non-human interaction. It's important to understand that with the constant connection these devices are generating big data regardless of the user interface. The action produces a filtered connection to control our behavior. Therefore, the only way to restore a natural unfiltered interface is by instilling the Unhackable mindset.

Spyware

Spyware is always being developed and used to spy on individuals. Recently, spyware "used on an Arab human rights activist led Apple to issue a global upgrade for several of its products… Vulnerabilities in Apple's operation system were exploited by an organization…that allegedly used the three newly discovered weaknesses in iOS, Apple's operating system, to take control of iPhone devices, including being allowed to read text messages and emails, track calls and have access to the phone owner's contact lists."[15]

The sophisticated spyware can even record sounds made around the phone, trace the user's location, and collect passwords put in the system. Now, Apple recommends to always download the latest version of iOS to protect yourself against security exploits.[16]

While downloading updates can be one security step, it isn't enough to keep you safe. Security systems are always needing updates and spyware is constantly being developed that exploits your computer and phone's weaknesses. Don't depend entirely on the tech company's updates to protect you.

Identity Theft Epidemic

Identity theft has become a huge security concern in recent years. The breach ruins lives as those who have been attacked suffer from a horrible invasion of their privacy. They may also deal with financial concerns as the thief uses their personal accounts or sets up new ones in their name. Invasive attacks leave victims vulnerable and afraid, and they are only getting worse. The Bureau of Justice defines identity theft as the:

- unauthorized use or attempted use of an existing account
- unauthorized use or attempted use of personal information to open a new account
- misuse of personal information for a fraudulent purpose.[17]

Instances are very common; many users have fallen victim to the scams. Unfortunately, the hackers gained access to their identities and bank accounts. One client comes to mind...

Stephen, a travel advisor, received a call from someone claiming to be from Microsoft. The person declared he was calling about his computer repair. Ultimately, the vicious hacker was able to lock the user out of his computer and gain access to his banking accounts. Stephen was left violated and exposed to the hackers' demands.

A distraught Stephen called to explain the situation. Thankfully, with some work, we were able to release his technology, money, and identity. He learned a valuable lesson and understands the process to protect his valuable private information. Stephen is cautious about Internet downloads and ignores similar calls. The incident left him empowered to make smarter decisions to avoid identity theft.

The Bureau of Justice Statistics presented this alarming information:

> About 7% of persons age 16 or older were victims of identity theft in 2014, similar to findings in 2012.
>
> The majority of identity theft victims (86%) experienced the fraudulent use of existing account information, such as credit card or bank account information.
>
> The number of elderly victims of identity theft increased from 2.1 million in 2012 to 2.6 million in 2014.[18]

Thieves are also stealing medical identities and using that information for their own gains.

While there is no comprehensive, federal accounting of this "disease," a survey of 2,000 U.S. consumers released by the firm Accenture gives some dimensions to the problem. Accenture found that 26 percent of U.S. consumers have had their personal medical information stolen from technology systems. Of those, around half (50 percent) were victims of medical identity theft. Evidence of this epidemic can be found in your local paper, where the police logs will document a steady stream of complaints to local authorities about online fraud, credit card fraud, malware attacks and the like.[19]

While we cannot control all the information that is released into various systems, we can limit the amount of content data online. Learn the proper protocols to protect yourself.

Spying

Modern electronic devices have extensive access and permission to our privacy. Users must educate themselves to enact security protocols to prevent an intentional campaign against them.

Even the most popular, up-to-date phones and computers are being designed in ways that invade our privacy. Now, phones have cameras and microphones that are always running. Malicious tools, such as Pegasus, can easily hack iPhones through something as simple as a text message. "It allows an attacker to install sophisticated spying tools that can activate a person's camera and microphone, track their movements, and log all messages."[20]

Recently, WikiLeaks published secret documents that portray the CIA as a "powerful hacking organization." A CNN report on the documents identified many of the ways the CIA can hack people, governments, and people:

> Phones, TVs spy on you: A team within the CIA developed spy software that infects Samsung smart TVs--placing televisions on a "fake off" mode that still listens to conversations and sends them back to American spies, WikiLeaks claimed. The program, called "Weeping Angel," was created with the help of the British spying agency MI5, it said.
>
> WikiLeaks reports that another team within the CIA built hacking tools that can remotely control iPhones, iPads and Android devices--secretly taking video from the camera, listening with the microphone, and tracking your location.[21]

Ex-National Security Agency spy Edward Snowden revealed that "Government spies are listening in on phone calls, collecting emails and tapping into people's Web cams."[22] CNN also interviewed former members of the CIA, Navy SEALs, and consultants to the U.S. military's cyber warfare team regarding the NSA's ability to listen to your conversations or follow your location. "Government spies can set up their own miniature cell network tower. Your phone automatically connects to it... Spies could keep your phone on standby and just use the microphone--or send pings announcing your location."[23] While these hacks are supposed to only be used against terrorists, the

NSA uses this surveillance for a variety of reasons in the United States.

Federal websites also use unauthorized software to gather information. *ABC News* brought attention to this particular invasion of privacy after a government investigation uncovered some concerning findings regarding our privacy. Sen. Fred Thompson, R-Tenn., chairman of the Senate Governmental Affairs Committee, released a report which found, "64 federal Websites that used unauthorized files that allowed them to track the browsing and buying habits of Internet users." This data is gathered through "cookies," which are "small software files that allow an Internet site to identify a specific computer that logs on to the site. Cookies can make browsing more convenient by letting sites distinguish user preferences, but the device has been attacked as an intrusion on privacy because they can track the kinds of Web sites frequented by a specific computer."[24] Although cookies are legal and used by various websites, they still create an invasion of privacy, and there are policies in place to prevent the government from using them in this way.

Wikileaks has also revealed multiple ways that Big Brother can hack into your system.

"'Year Zero' introduces the scope and direction of the CIA's global covert hacking program, its malware arsenal and dozens of 'zero day' weaponized exploits against a wide range of U.S. and European company products, including Apple's iPhone, Google's Android and Microsoft's Windows and even Samsung TVs, which are turned into covert microphones," WikiLeaks wrote in a press release announcing the trove of documents.

According to the release, the CIA can also "bypass the encryption of WhatsApp, Signal, Telegram, Wiebo, Confide and Cloackman."[25] Unfortunately, this kind of technology continues to be used in unauthorized ways.

Google and Other Search Engines

Search engines store history of every interaction unless users configure access and permissions in their account settings.

CNBC's Todd Haselton discovered Google knew his name, gender, birthdate, personal cell phone number, websites visited, when he turned his bedroom lights off, exactly where he's been over the last couple of years, and every time he has used his voice to interact with Google.[26] Apparently, all of this was built to deliver accurate search results, but at what sacrifice?

What sacrifices are we willing to make and what risks are we willing to take? "The recordings can function as a kind of diary, reminding you of the various places and situations that you and your phone have been in. But it's also a reminder of just how much information is collected about you, and how intimate that information can be."[27] Think about how much information is collected about you and how intimate that information can be. Will you think twice now that you know data brokers are purchasing information stored on the Internet to sell to other companies to use and hackers to steal?

Google keeps track of everything you search, all of your emails (even those unsent) on Gmail, everything you share on Google+, every website you visit on Chrome, Firefox, or Safari, everything you watch on YouTube, and everything you do on your Android device. Google also uses Google Analytics, AdSense, AdWords, Widgets, Street View, Google Glass, and Google Maps to track you.[28] Using this data alone can steer results in a particular direction and effectively influence the entire world, because Google can predict the future based on trends. Don't forget that search history and location data is retrieved as Google conducts its query. Not only does Google have your search engine information and your interests, but even your precise location at any given moment.

Recently, Google presented a red question mark in Maps on a client's phone after a new update. It prompted him to enter his home and work address. He made the mistake of entering the information without thinking too much about it. While he was driving, his phone beeped to let him know he was twelve minutes away from his house. With Maps, Google knows where you live and where you are at all times. "Using Google Maps on a smartphone requires reporting your location to Google, and this continues even when the Maps application isn't running."[29] Beyond the concerns about what search engines could do with the security of historical information, it is also dangerous because hackers could get this information and know where you are and what you are doing.

Facebook and Other Popular Social Media Sites

Social media sites, such as Facebook, can be great tools for connecting with friends and family and building community online. However, it can be difficult to maintain your privacy with social media.

Safety should always be your main priority on the Internet, so make sure you have secured your account before sharing content on social media.

When reviewing the app's permissions group on an Android, you can see that Facebook Messenger has permission to record you.[30] Additionally:

> The TOS (Terms of Service) also authorizes Facebook to take videos and pictures using the phone's camera at any time without permission, as well as directly calling numbers, again without permission, that could incur charges...Since the vast majority of people will agree to these terms without even reading them, cellphone users are agreeing to let Facebook monitor them 24/7, green lighting the kind of open ended wiretap that would make even the NSA jealous.[31]

Why would Facebook need these permissions? What are they doing with the information they are gathering?

As a user of Messenger, you can quickly recognize that content from your Messenger communications is being shared with advertisers. Mention to a family member that you are looking into buying a new car, and all of a sudden, you'll get ads for the make and models you discussed. If you write to a friend about an upcoming vacation, expect to see some ads on Facebook about hotels and sightseeing in that area. Additionally, Facebook has incorporated ads into its Messenger app. "In a world populated by ad-blocking through browsers, segregated apps such as Facebook's and Twitter's have remained isolated from the trend, and a place to guarantee eyeballs on ads for marketers, aided by Facebook's wealth of user data and advanced targeting."[32]

In addition to invading your privacy, Facebook influences users through social experiments and explores ways to control emotions through what is shared. NPR shared findings from a scientific paper that shows how Facebook conducted a study on changing users' moods based on newsfeed content:

> The social media company altered the news feeds (the main page users land on for a stream of updates from friends) of nearly 700,000 users. Feeds were changed to reflect more "positive" or "negative" content, to determine if seeing more sad messages makes a person sadder. The bottom line is news feeds were tweaked without warning because Facebook users agreed to the social giant's general terms of data use, and researchers tracked emotional responses of test subjects by judging any subsequent changes in their use of language.[33]

All of this is used to keep you online, where advertisers and hackers can easily target you.

One huge data scandal that has been brought to light has to do with Cambridge Analytica, a UK-based political-data company, which illegally used users' data to target political ads. "The data, a portion of which was viewed by The New York Times, included details on users' identities, friend networks and "likes." The idea was to map personality traits based on what people had liked on Facebook, and then use that information to target audiences with digital ads."[34] Not only is your data being used to sell you things, it is also being used to sell ideas, control your emotions, and influence your actions.

Make sure to manage your social media content, even if your profile is set to private. Facebook still shares your private information with advertising companies. "The issue affects tens of millions of Facebook app users, including people who set their profiles to Facebook's strictest privacy settings."[35]

It is mind-blowing just how much information Facebook gathers on its users. CNBC's Todd Haselton shares how to find all the data Facebook stores about you:

> Facebook stores almost every single interaction you've had with the social network since you joined, including every time you've logged in, ads you've clicked, events you've been invited to, a list of the people you follow, your friends, your hometown, every time you've sent or received a message, every single status update and more…[36]

You can download your own archive of this data from Facebook. Here's how:

> Go to Facebook.com/settings
> Tap "Download a copy of your Facebook data."
> Tap "Download Archive."
> It might take a few minutes, but Facebook will alert you when your archive is ready.
> When it is, click "Download Archive" again, and a zip file will download to your computer.

Browse through that archive by opening each file inside the folder.[37]

Be very cautious about the information you put on social media sites. As long as you secure your account, you can continue to use social media. However, be very cautious about the information you put on these sites.

Accidental Hackers

As technology changes, it can be difficult to know what steps to take to protect ourselves from user error and costly, uninformed choices. Many parents have learned the hard way about in-app purchases and password protection workarounds due to curious kids using technology without truly understanding the dangers.

USA Today reported on a girl who used her sleeping mother's thumbprint to make online purchases. The six-year-old unlocked the Amazon app with her mother's fingerprint and "then proceeded to order $250 worth of Pokemon presents for herself." As humorous as this hacker story is, "it shows just how easy it is to foil security features... 'I mean, really, if a six-year-old can beat it, just how secure is it?'"[38] She isn't the only child to surprise her parents with unapproved purchases from their phones. Many app developers make it all too easy for children to spend hundreds of dollars on in-app purchases.

Another six-year-old quickly spent well over $100 on Smurf Village credits in a game his father bought him. When the dad saw the charges, he immediately thought he'd been hacked, but quickly realized the "hacker" was his son. "I contacted Apple and discovered I wasn't the only naive parent in the world. This is a common occurrence and Apple refused to issue any kind of refund. Needless to say that I have now disabled all in-app purchases on my devices. Lesson learned."[39] These can be very expensive lessons to learn and app designers and developers know how to make these purchases appealing and easy for children growing up in a tech-saturated world.

Every user must become a Chief Digital Officer (CDO). In the process, you will become your own data boss and begin your digital transformation.

5

CYBER DANGERS OF THE FUTURE

"Cyberwar is the battlefield of now."
—Geoff Livingston

The future of technology is unknown; however, there are some distressing trends developing that indicate technology will be playing an even greater role in our lives. As wonderful as innovation can be, the security issues have not been resolved. Corporations and hackers are gaining more access to personal data as the Internet evolves. It will continue to remain unchecked until users take back control of their online data.

Already, artificial intelligence is infiltrating our homes, cars, and jobs, and even gaining citizenship! Hanson Robotics' AI Sophia has been granted citizenship by Saudi Arabia. At the Future Investment Initiative conference, Sophia stated, "Thank you to the Kingdom of Saudi Arabia. I am very honored and proud for this unique distinction. It is historic to be the first robot in the world to be recognized with citizenship."[1] Although there are many future possibilities on the horizon, we need to continue to be aware of how these advances in technology put us in danger.

"Smart Homes"

In the near future, artificial intelligence will play a much greater role in our lives, especially as the "smart home" trend continues to rise. The Internet connects people as never before, but it also creates tension within the communal existence.

The more technology we have in our homes, the more our privacy is threatened. While these innovations might appear to make your life easier, it also makes it easier for influencers to manipulate users. Since we have explained the seriousness of our online connections, why are we moving forward and putting our families and businesses at risk?

One of the first robots to enter many people's homes is the Roomba, the robot vacuum that helps ensure your floors are always clean. While it may seem like this type of technology couldn't be a danger, Roombas have been gathering data as they move around the floors of our homes. According to the Gizmodo article "Roomba's Next Big Step Is Selling Maps of Your Home to the Highest Bidder," "it knows the floor plan of your home, the basic shape of everything on your floor, what areas require the most maintenance, and how often you require cleaning cycles, along with many other data points."[2] Now, those data points are going to be sold to the highest bidder.

It all goes back to the invention of inbound marketing. As long as corporations have the ability to bring the consumer to the product, the invasion will continue. You might believe that your floor plan data isn't that valuable, but it can be used to sell you more and more technology, which continues to gather more and more data on you and your home.

If a company like Amazon, for example, wanted to improve its Echo smart speaker, the Roomba's mapping info could certainly help out. Spatial mapping could improve audio performance by taking advantage of the room's acoustics. Do you have a large room that's practically empty? Targeted furniture ads might be quite effective. The laser and camera sensors would paint a nice portrait for lighting needs that would factor into smart lights that adjust in real time. Smart AC units could better control airflow. And additional sensors added in the future would gather even more data from this live-in double agent.[3]

The threat is not just a threat to marketing. Companies gather data, then share the information for additional profit. Think about how dangerous the floor plan of your home could be in the hands of a hacker. All of this from one little robo-vac!

The future of "smart home" technology goes way beyond Roombas. Mark Zuckerberg, the Facebook founder, is creating a program called Jarvis. With Jarvis, you can walk into your house and tell it to do everything for you. "Zuckerberg designed his personalized home automation network to make accomplishing such tasks as playing music or dimming the lights easy while also allowing the AI to personally learn his 'tastes and patterns.'"[4] With this type of technology, Artificial Intelligence is tracking all of your requests and gathering a lot of data. These companies are making a lot of money, but have they thought about people and their lives and data?

While Jarvis is a work in progress, there are already multiple AI home assistants on the market. Amazon Echo's Alexa and Google's Home are two of the most popular. We are beginning to interact with Artificial Intelligence as if it is a person, not a device. Many are becoming a little too emotionally attached. "Over 100,000 people say good morning to Alexa each day, and 250,000 have asked for its hand in marriage."[5] These assistants may seem like they are improving our quality of life; however, we can't forget that they are monitoring us and storing that data every day that we have them in our homes.

In addition to privacy concerns, the increased use of smart technology takes control and decision making out of our hands.

In addition to playing music or adjusting lights, Jarvis goes much further in taking over the everyday tasks in our lives. It can "open the front gate for friends, make toast, and even wake up their [Zuckerberg's] one-year-old daughter Max with Mandarin lessons."[6] While this may seem appealing, as AIs take on these tasks, we will end up in a world where nobody knows how to do anything anymore.

"Smart Home" technologies, such as Google Home, also control the information we receive. Our ability to access accurate information and use our critical thinking skills is quickly fading. Google Home has been accused of pulling "from sites sharing fake news, propaganda and simple lies. Worse, it can result in the Google Home reading the same statements as fact, without even the presence of the other search results to provide much needed contextual clues that the answers might be misleading."[7] Do you want something like Jarvis or Alexa controlling your life or would you rather use the mind you were born with that gives you free will and choice?

"Products within the home are increasingly connected to make things safer, easier, and more productive." The uses of the Internet of Things are everywhere and growing in connected healthcare, connected hospitality, connected cities, and connected homes. Their reason for having connected homes is that heating, security, and maintenance will improve the quality of life for homeowners.[8] So now, hackers can get to our equipment and access our homes. We need to wake up before it's too late.

"Smart Cars"

In addition to "Smart homes," we are now designing "Smart" self-driving cars. This is another step toward giving up our control and free-will. With the introduction of Tesla's "autopilot" system, drivers are putting themselves in danger by overly relying on their car's advanced tech. Los Altos Planning

Commission chair Alexander Samek was involved in a police chase when he fell asleep at the wheel while his Tesla was operating on autopilot. Another driver received a DUI after falling asleep at the wheel while intoxicated. A Tesla on autopilot was also involved in a fatal crash where the car sped up moments before the crash. California Highway Patrol Public Information Officer Art Montiel stated, "It's great that we have this technology; however, we need to remind people that... even though this technology is available, they need to make sure they know they are responsible for maintaining control of the vehicle."[9]

He is not alone in this sentiment. As more car companies bring autopilot capabilities to their cars, there is a concern about drivers overly relying on technology. Andy Christensen, lead engineer on Nissan's ProPilot Assist has stated, "It's on us to educate people about what's allowable and what's not allowable. We don't want drivers to be overly confident. (The tech) is there to assist you, it's not driving for you."[10] Not only are smart cars unreliable on autopilot, they also open us up to more attacks from hackers. For example, a smart car's location can be tracked even if the car is off.

In a Phys.org article "The Cybersecurity Risk of Self-Driving Cars," Engin Kirda, a systems, software, and network security expert who holds joint appointments in the College of Computer and Information Science and the College of Engineering at Northeastern University, shared his concerns about self-driving cars:

> In principle, any computerized system that has an interface to the outside world is potentially hackable. Any computer scientist knows that it is very difficult to create software without any bugs—especially when the software is very complex. Bugs may sometimes be security vulnerabilities, and may be exploitable. Hence, very complex systems such as self-driving cars might contain vulnerabilities that may be potentially

exploited by hackers, or may rely on sensors for making decisions that may be tricked by hackers. For example, a road sign that looks like a stop sign to a human might be constructed to look like a different sign to the car.[11]

Be extremely cautious about embracing this new "smart" technology. Unvetted innovation increases the risk substantially. Innovation is infectious as any form of technology can be hacked, so it's wise to set up, configure, and install all Internet devices with the proper access and permission controls in place.

Tech Toys

Another danger that we bring into our homes are the toys we give to our children. We are at a crossroads for our children. Every toy is connected to the Internet, yet nothing is secure. More and more tech toys are being put into the marketplace each year, but many of them "pose an imminent and immediate threat to the safety and security of children." Two popular toys that "listen" and talk to children, the My Friend Cayla doll and the i-Que robot, "capture a user's voice without providing adequate notice or obtaining verified parental consent. An insecure Bluetooth connection, meanwhile, allows anyone to eavesdrop on children and their dolls, creating a risk of 'predatory stalking or physical danger.'"[12]

There are also concerns that some countries could be using these toys to eavesdrop. "According to a new complaint filed with the Federal Trade Commission from a coalition of consumer privacy advocates including the Electronic Privacy Information Center (EPIC). The toys allegedly send recordings to speech-to-text software company Nuance Communications, which the complaint notes has contracts with military and law enforcement agencies."[13] Nothing is safe; not even the toys you give to your children to bring them joy. Parents need to be more informed about tech security in order to protect their children.

Robots in the Workplace

While Artificial Intelligence is playing more of a role in our homes, robots are also moving into the workforce. The digital business world has sparked a shift in the relationship between man and machine. Everyone is now focusing on the idea of human-machine cooperation and growth. Machines are taking a more active role in enhancing human endeavors. Machines are more connected than ever before and they have an increased ability to supplement human jobs and to reduce the cost of operations. According to Gartner, "By 2018, more than 3 million workers globally will be supervised by a 'roboboss.' Robobosses will increasingly make decisions that previously could only have been made by human managers."[14]

Not only will robots be overseeing our work, they will also be replacing our jobs. "Researchers quizzed business leaders on how they see automation and artificial intelligence affecting their industry over the coming years. Over 20 percent of employers in finance, accounting, transportation, and distribution stated that they expect more than 30 percent of jobs in the field to be automated by 2027."[15] Many CEOs and consultants at *Fortune* Brainstorm Tech claim that every type of job will be affected by artificial intelligence in the near future. According to *Fortune's* article about Brainstorm Tech, "These are the Jobs that Artificial Intelligence Will Eliminate First,"[16] there are certain "jobs that almost certainly will disappear as AI and machine learning technologies continue to evolve and become more prevalent. This includes things like drivers (thanks to autonomous vehicles), lower-skilled manufacturing jobs (humans out, robots in), and certain research functions (paralegals, etc.)."

The increased use of computing machines in decision making is extending into the realm of financial choices with increasing consistency. One must think about how many of the economic decisions we make will be supported by, then automated through, digital technologies.

Soon, even massage therapists, writers, and investment professionals will see their jobs threatened by robots. One CEO shared why his company is using a robot named Burt; "our engineers build software so every one of our 75,000 clients get the benefit of Burt, but each of them gets the exact same quality of service from a technology that is available 24/7 and doesn't get emotional." What happens when the robot can no longer solve the issue, or there is a bug in the code? Do we really want to live in a world lacking human interactions?

The innovation of replacing the human brain with artificial intelligence carries some serious red flags. While I acknowledge the importance of digital technology and love the opportunities it brings forth to all sectors and users, the Unhackable mindset will draw a demarcation line in the sand. This is where the business model has gotten out of control and is now shifting from man to machine.

We should be aware that humans must be the designers of Artificial Intelligence; never the Artificial Intelligence designing themselves. The lure of going beyond just programming a machine and losing all human elements is a real and present danger. As artificial intelligence evolves, it is becoming more than just a box of circuits. The influence is destroying the natural human connection. Humans were created to look beyond the box container with creativity and imagination. Just look at the Universe; is there a limit? It continues to expand, while AI decreases our abilities. As an experiment, close your eyes and imagine the Statue of Liberty; can a computer look beyond its framework? Will it ever be able to create a neutral idea? It is always adjusting from its original programming, leading humans to a filtered isolation that was never in our nature.

Our technology is taking over one step at a time. Just because the use of AI will reduce costs, save time, and improve accuracy, does that make it right to be in control of our thought processes? Is anyone thinking about privacy and security, especially in this cyber warfare era? How can we consider AI to this level when we cannot control the present chaotic connected environment enveloping society?

Bitcoin

The present movement is trying to move us away from traditional currency and move toward using bitcoin. According to Bitcoin.org:

> Bitcoin uses peer-to-peer technology to operate with no central authority or banks; managing transactions and the issuing of bitcoins is carried out collectively by the network. Bitcoin is open-source; its design is public, nobody owns or controls Bitcoin and everyone can take part. Through many of its unique properties, Bitcoin allows exciting uses that could not be covered by any previous payment system.[17]

The concept of using cryptocurrency as a replacement for paper money has its place in the future of technology, but only when we have resolved the issue of cybersecurity. Offline federal banking has established regulations that openly protect account holders. Bitcoin was developed by an unknown person or company; the only one claiming responsibility is someone named Satoshi Nakamoto. The currency has no central bank or administrator and the monies can be sent from one person to another across an unsecured network. Hackers, as we have proven in the previous chapter, use Bitcoin as payment in Ransomware scams. Do you want to invest your money in such an unscrupulous environment? Bitcoin is an alternate option of currency that brings forth some major concerns about the validity to the money. Not all innovations are meant to make our lives easier. We should try to work within the boundaries already set in place.

Chips

As unbelievable as it may seem, companies are already implanting chips in their employees. *The New York Times* recently published an article about the first company in the United States that is embracing chips. In "Microchip Implants for Employees? One Company Says Yes," the company, Three Square Market, says these chips will be used for easier building access and payments for cafeteria food. Many embrace chip implants as progressive and cutting-edge, but there are many security concerns about this new use of technology.

This type of technology could easily be used for other purposes that invade employees' privacy. According to Alessandro Acquisti, a professor of information technology and public policy at Carnegie Mellon University's Heinz College, "a microchip implanted today to allow for easy building access and payments could, in theory, be used later in more invasive ways: to track the length of employees' bathroom or lunch breaks, for instance, without their consent or even their knowledge."

Acquisti wisely pointed out that "companies often claim that these chips are secure and encrypted...but 'encrypted' is a pretty vague term...which could include anything from a truly secure product to something that is easily hackable."[18] Nothing is truly secure on the Internet, so it is very likely that hackers will target these chips in the future. Easier isn't always better.

Chips aren't the only way technology is crossing over to be a part of our bodies. Right now, "Tech Tats" are being put on people, even children, to monitor health. These devices are mounted on the skin and integrate electronics with the human body. It can monitor everything that your doctor would check during an annual physical. The goal is to make them as cheap and accessible as band-aids.[19] Like a lot of the technology discussed in this book, it may seem like a good idea, but we don't know how far it will go and need to be cautious about connecting to technology in this way.

Thought Crimes

The Internet has already greatly increased the amount of personal data that others have access to—our likes, dislikes, location, activity, the list goes on and on. But what about our thoughts? Do you want your technology to read your mind? Thanks to artificial intelligence and big data, technology can decode your brain activity and reveal what you're thinking and feeling. Will our thoughts always stay private and protected? Due to our usage and big data, patterns can be uncovered very easily. Technological advances and startling trends in some workforces suggest that our thoughts and emotions will someday be more data for others to gather and use against us. Bioethicist, lawyer, and philosopher Nita Farahany presented a TED Talk exploring the dangers of mind-reading technology. EEG devices already record brain activity and can identify moods and simple words and phrases. This technology can be used for good, such as helping an epileptic know when they are going to have a seizure, but Farahany fears that this technology could be used to invade our minds. "I worry that we will voluntarily or involuntarily give up our last bastion of freedom, our mental privacy. That we will trade our brain activity for rebates or discounts on insurance, or free access to social-media accounts... or even to keep our jobs." She shares that train drivers in China are already required to wear EEG sensors. "Workers are even sent home if their brains show less-than-stellar concentration on their jobs, or emotional agitation."

Her fears are even greater than just being sent home from work. "If our brains are just as subject to data tracking and aggregation as our financial records and transactions, if our brains can be hacked and tracked like our online activities, our mobile phones and applications, then we're on the brink of a dangerous threat to our collective humanity."[20]

What will this mean for our already violated sense of privacy? Do we want a society where people are arrested for merely thinking about committing a crime (like in *Minority Report*) and private interests sell our brain data? It's time to create a case for the right to cognitive liberty that protects our freedom of thought and self-determination.

Cyber Warfare

Imagine that a switch goes off, and you have no access to the Internet. How disabled is your life? How disabled is your business? After a massive Internet outage on the East Coast due to a DDoS attack, dozens of sites and services were disrupted. According to *Wired*, this is not a new type of attack, and they aren't easy to stop. "Initial reports indicate that the attack was part of a genre of DDoS that infects Internet of Things devices (think webcams, DVRs, routers, etc.) all over the world with malware. Once infected, those Internet-connected devices become part of a botnet army, driving malicious traffic toward a given target."[21] We live in a world where this type of attack will happen again, and in the future, the consequences could be much worse. Most of these attacks are resolved quickly, but there are no guarantees. We assume that the Internet will always be there for us, but cyberattacks like this can take sites down all over the world.

Cyberwarfare can do more than impact computers. Stuxnet, the world's first digital weapon, is "unlike any other virus or worm that came before. Rather than simply hijacking targeted computers or stealing information from them, it escaped the digital realm to wreak physical destruction on equipment the computers controlled."[22] The documentary *Zero Days* explains the history of its discovery.

One dramatic sequence shows how the Symantec team managed to drive home Stuxnet's ability to wreak real-world havoc: they programmed a Siemens PLC to inflate a balloon, then infected the PC it was controlled by with Stuxnet. The results were dramatic: despite only being programmed to inflate the balloon for five seconds, the controller kept pumping air until it burst.[23]

We're in a world of cyber warfare. Our enemies are using the Internet for malicious attacks that can wreak havoc on our infrastructure and defenses while gaining access to top secret information that can be used against us. A Reuters Special Report "Government in Cyber Fight but Can't Keep Up," states:

> In recent months hackers have broken into the SecurID tokens used by millions of people, targeting data from defense contractors Lockheed Martin, L3 and almost certainly others; launched a sophisticated strike on the International Monetary Fund; and breached digital barriers to grab account information from Sony, Google, Citigroup and a long list of others.
>
> The latest high-profile victims were the public websites of the CIA and the U.S. Senate—whose committees are drafting legislation to improve coordination of cyber defenses.[24]

In the report, Jim Lewis, a cyber expert with the Center for Strategic and International Studies think tank states "the network is so deeply flawed that it can't be secured." It's all about being informed; it's about knowing your enemy.

Former FBI director James Comey admits to serious flaws in cybersecurity in the United States:

> He said the FBI is renewing a focus on the challenges posed by encryption. He said there should be a balance between privacy and the FBI's ability to lawfully access information. He also said the FBI needs to recruit talented computer personnel who might otherwise go to work for Apple or Google. "The cyberthreats we face are enormous. I don't know if we can stay ahead of them. And I think to say otherwise would be hubris," Comey said.
>
> "We need to ensure that cybersecurity is a priority for every enterprise in the United States at all levels. We need to get better and faster at sharing information in the appropriate ways. We

need to make sure we have the right people on board to help fight that threat, and we need to build trust between the government and the private sector," he said.[25]

The advancement of innovation at this rapid pace has created security holes and weaknesses that far outweigh the benefits and advantages. Users are not thinking about how technology is invading their privacy. We have been targeted through the ease and marketing of online storage and availability for our data as other options are taken away or made much harder to find. How dangerous is that? We're moving data from our own private computers and putting it online for anyone to access at any time.

War in Space

Our enemies have access to content data stored online at any time. However, that isn't the only way they can use technology against us. Because of our dependency, the wars of the future will be very different from anything we've seen previously. The most powerful and destructive way to attack us is through space. As a civilization, we rely so much on technology that society will be crippled in the event of a cyberattack that knocks out technology.

In the CNN Special Report, *War in Space—The Next Battlefield*[26], military officials reveal just how detrimental attacks on our satellites could be for our military and our day-to-day lives. We think about GPS technology being used for maps, but the military depends on it. These satellites also guide clocks, ATMs, gas pumps, stop lights, hospital records and procedures, and financial transactions. Communication satellites are also heavily depended on by the military. Without these satellites, troops on the ground would be in danger without any commands. Our military and civilians would be dragged back in time and become incredibly vulnerable if this technology was attacked.

Right now, the satellites we've come to depend on are unarmed because we didn't think they were at risk when they were sent out into space. However, laser weapons can now temporarily blank or disable space assets, kamikaze satellites can destroy other satellites, and ground-based missiles can reach space to destroy important satellites.

When we lose power because of a hack, what will you do then? When that switch goes off, what are our children going to know? This is a serious threat. When the time comes, we can either be devastated by the attack or prepped to continue without technology.

6

A GREATER THREAT

"Media exposure has become America's most widespread and serious addiction."
—George Barna

Loss of Identity

Since our lives are online, our identity depends on technology. The movie *The Net* foreshadowed the future of technology where it has the ability to completely eliminate a person's entire life, so that they never existed. Granted, it's a movie, but the reality is very apparent.

Internet usage is highly addictive. Often, we see friends out together but absorbed in their phones, families on their devices rather than focused on one another, concert-goers focused on their online persona rather than concentrating on the moment.

As dangerous as it is to have our data gathered, stolen, and used against us, the greater threat of technology is the loss of self, caused by the intentional algorithms created to keep us absorbed and occupied. Our vision for our lives has been clouded due to the massive amounts of time wasted on our digital journey.

In the film *Captivated: Finding Freedom in a Media Captive Culture,* everyday citizens share their journeys from tech addicts to advocates for tech-free living. Writer, producer, and co-director Phillip Telfer observes:

> It's not the subject people have a problem with, it's the steps that need to be taken to find freedom that many people are uneasy with. Maybe uneasy is too soft of a word. I have great hope that things can change for better regarding media discernment and making wise media and entertainment choices, but that will take some action in our personal lives, in our homes, in our communities, at our workplaces, and in our churches. And the best place to start is to consider this area in your own life. It's easier to think about someone else who may need to make changes in their media habits, but each one of us needs to take inventory of how our lives, our relationships, and our walk with God are impacted by the digital age.[1]

Erik Engstrom read Neil Postman's *Amusing Ourselves to Death* and was inspired to change the way his family uses technology. He decided to start this positive change by having his family go on a one-month media fast. Once they began, it was pretty shocking to them how addicted they really were to their technological devices. As they spent time away from technology, they became more connected to what they really valued in life: one another. Wife Karen shared that the children "loved the extra time of playing basketball with dad or playing cards with me or just working on extra things." In discussing the family's media fast, their teenage daughter Keilah said, "This has been so good in our lives, and it hasn't been boring at all!" Learn more about the Engstroms and other families who have made changes at www.captivatedthemovie.com.

Children and Technology

How dependent are your children on technology? Could they function if their technology was taken away? Our children rely so much on technology that they don't know how to socialize and interact with others in person. The use of IoT-enabled devices is growing dramatically, which reduces the amount of in-person interaction users experience. "IoT is a scenario in which objects, animals or people are able to transfer data over a network and talk to each other without requiring human-to-human or human-to-computer interaction."[2] According to UCLA research presented in the NPR article "Kids and Screen Time: What Does the Research Say?," "kids are spending more time than ever in front of screens, and it may be inhibiting their ability to recognize emotions." One UCLA professor, Patricia Greenfield, says, "our species evolved in an environment where there was only face-to-face interaction. Since we were adapted to that environment, it's likely that our skills depend on that environment. If we reduce face-to-face interaction drastically, it's not surprising that the social skills would also get reduced." The article also pointed out that "some research suggests that screen time can have lots of negative effects on kids, ranging from childhood obesity and irregular sleep patterns to social and/or behavioral issues."[3]

The heavy use of technology even hurts young children's motor skills. Dr. Radesky, a clinical instructor in Developmental-Behavioral Pediatrics at Boston University, shared some of the concerns that came out of recent Boston University School of Medicine research:

> Heavy device use during young childhood could interfere with development of empathy, social and problem solving skills that are typically obtained by exploring, unstructured play and interacting with friends...These devices also may replace the hands-on activities important for the development of

sensorimotor and visual-motor skills, which are important for the learning and application of maths and science.[4]

As more and more tech devices are invented and embraced by the masses, the majority of Silicon Valley quickly became aware of some of the dangers technology presents, especially its addictive qualities. In fact, they limit technology use in their homes. "In 2007, Gates, the former CEO of Microsoft, implemented a cap on screen time when his daughter started developing an unhealthy attachment to a video game. He also didn't let his kids get cell phones until they turned 14."[5] Steve Jobs didn't even give his children iPads. "Walter Isaacson, the author of *Steve Jobs*, shared with New York Times reporter Nick Bilton, 'Every evening Steve made a point of having dinner at the big long table in their kitchen, discussing books and history and a variety of things,' he said. 'No one ever pulled out an iPad or computer. The kids did not seem addicted at all to devices.'"[6] Chamath Palihapitiya does not allow his children to have any screen time, sharing with CBS This Morning, "I want children who can make eye contact. I want children who know how to resolve conflicts with their peers. I want children who understand the dynamics of interpersonal relationships that are physical and tactile. I do not want children who only know how to interface with the world through a screen."[7]

Melinda Gates shares her concerns and suggestions with parents in a piece for the Washington Post: "I spent my career in technology. I wasn't prepared for its effect on my kids." Even though the Gates strongly limited technology, they still struggled to keep up with the latest innovations and keep track of the impact it would have on their children. She shares, "I spent my career at Microsoft trying to imagine what technology could do, and still I wasn't prepared for smartphones and social media. Like many parents with children my kids' age, I didn't understand how they would transform the way my kids grew up—and the way I wanted to parent. I'm still trying to catch up."[8]

Since children's use of technology is increasing so quickly over a short period of time, we don't even know the long-term consequences of their overuse of technology. Children are the future. Young kids don't have the social skills to know what information is put online that can hurt them. Once a post, comment, or photo is uploaded, it lasts forever online and is available to whomever in the outside world is prying into their lives. As innovation advances, children spend an inordinate amount of time online. It is altering their personalities. As technology is used to do more activities for our youth, the trend is alarming. We must ask ourselves what can be done to change this situation to make a safer world for humanity and better the lives of our children?

The time has come for parents to take an active role in educating their children on the dangers of technology. In my opinion, an unhealthy compulsion is taking place regarding a technology dependence. Parents need to be aware of the dangers and educate their kids. For more information about how technology is affecting our children, watch *Screenagers*, a film that "probes into the vulnerable corners of family life, including the director's own, and depicts messy struggles, over social media, video games, academics and Internet dependence. Through surprising insights from authors and brain scientists solutions emerge on how we can empower kids to best navigate the digital world."[9] Beyond the documentary itself, the website (www.screenagersmovie.com) provides parents with useful resources, such as a Family Screen Time Agreement Template.[10] Use this to help guide you as you work to find a balance between screen time and other activities with your family.

In addition to this agreement, you might also want to consider using a cell phone contract with your child. Verywell Family has a contract that you can use, which includes safety points such as "I understand that having a cell phone can be helpful in an emergency, but I know that I must still practice good judgment and make good choices that will keep me out of trouble or out of danger" and "I promise I will alert my parents when I receive suspicious or alarming phone calls or text messages from people I don't know."[11]

Many parents are so engrossed in their own social media interactions and texting with other people that they are not aware of what's happening all around them. Much of society has become so mesmerized by this self-made virtual reality that we have secluded ourselves from the physical world. The education of our children must come from face-to-face interactions with parents or guardians. One example might be to have family meetings and discuss concerns about the dangers of technology. However, meetings must include the children and their knowledge or questions about using these digital devices. A healthy plan would be to learn security measures with the children as a team.

Our children must know about the privacy and health concerns related to technology. The Internet is always being monitored in some fashion. Due to the addictive attributes of technology, time limits should be enacted upon children who have access to digital devices. The American Academy of Pediatrics' Healthy Children site provides a great tool for creating a family media plan. Learn more at: www.healthychildren.org/English/media/Pages/default.aspx

As a part of your family plan, you may decide to limit the use of technology during important family times, such as dinner. Common Sense Media's "#devicefreedinners" (www.commonsensemedia.org/device-free-dinner) is a movement that encourages and supports families limiting the use of technology in their homes.

When you put limitations on technology, make sure to replace that time with something else. Plan fun activities. Get outside. Dr. Jeff Myers, President of Summit Ministries shared this wisdom in the previously mentioned documentary

Captivated: Finding Freedom in a Media Captive Culture, "when you come home and say, 'Kids, we're gonna unplug, cause I've got all this new information here.' There will be absolute warfare in most households. Why? Because the media has actually fostered an addiction. I would recommend as a family that you plan out a week or two worth of activity, fun things that you can do outside. Just go play miniature golf. Go play frisbee golf, go play tennis, get some different things that you can do outside, but make sure that you're together doing those things outside."[12] We need to connect with our children, free from technology. It has to start today. We need to educate them correctly. Don't wait or it will be too late.

Relationships and Technology

Children aren't the only ones living in a world with limitations being created on physical human interaction. Many people struggle these days with personal conversations. Because of technology, most of our socializing happens online. Journalist and author Maggie Jackson shares her concerns about how technology affects relationships. "I'm most concerned that we are actually shifting for the worse the definition of what it means to be a human. If we are satisfied with the snippet relationship, with the faceless communication, even for our most intimate loved ones, then we stand the chance of robotizing relationships."[13]

We try to put the best versions of ourselves online and often lose our true selves in the process. The impact has a major effect on relationships. We use social media as a way of expressing our feelings and emotions; just look at the invention of emoticons. The issue is causing problems with communicating and people have become dependent on social media to validate our relationships. As O'Connor points out in her article "How Technology Is Controlling Our Relationships,"

We're living our lives for an audience, and dying for their approval. How many times have you heard phrases like "It's not official till it's Facebook official." Or "Ooh you posted a picture with them on Instagram so I guess it's getting pretty serious." Just listen to how insane that sounds…

We should not accept a heart eye emoji as a gesture of love, or a text back as someone putting in "effort." We need to return to connecting with people on real levels, and using technology just as a means of contact, not a device to control our lives.[14]

Social media can also have a negative impact on our health. It causes depression, leads to addiction, leads to isolation, inhibits physical interaction with others, hinders creativity, and decreases physical activity.[15]

Not only are there negative impacts on our health, but as you now know, the use of technology is also detrimental to our privacy and safety. Privacy settings are always changing and dynamic on websites. As we share more of our lives with friends and family online, we are also sharing our data with companies, foreign governments, and hackers who will use it for their own gain. According to the FBI's 2015 Internet Crime Report, "millions of people in the United States are victims of Internet crimes each year. Only an estimated 15 percent of the nation's fraud victims report their crimes to law enforcement. This is just a subset of the victims worldwide. Detection is the first piece of the puzzle and is the cornerstone of the larger Internet crime picture."[16]

Loss of Joy

While we often turn to technology to make us feel better through online shopping, gaming, and interacting, multiple studies show that those who spend more time online have a lower quality of life and struggle more with depression. Many teens rely on social media as their main form of interaction with their peers. They can become consumed by messaging apps like Snapchat, Kik, and WhatsApp and lose sight of what is happening around them. The Atlantic's "Have Smartphones Destroyed a Generation?" provides an in-depth look at how young people are suffering mentally and emotionally because of their reliance on technology:

> All screen activities are linked to less happiness, and all non-screen activities are linked to more happiness. Eighth-graders who spend 10 or more hours a week on social media are 56 percent more likely to say they're unhappy than those who devote less time to social media. Admittedly, 10 hours a week is a lot. But those who spend six to nine hours a week on social media are still 47 percent more likely to say they are unhappy than those who use social media even less. The opposite is true of in-person interactions. Those who spend an above-average amount of time with their friends in-person are 20 percent less likely to say they're unhappy than those who hang out for a below-average amount of time.[17]

Rather than being present in the moment and creating experiences in the real world, we are focused on finding the best Instagram filter to get likes from our followers. We are not finding joy and meaning in the actual experience, but are instead looking for virtual validation.

Teenagers aren't the only ones using social media more and more in a quest for connections that leave them feeling unhappy. As a professor of business and psychology, Adam Alter has been studying the effect of screens on our lives. He presents some of his findings in his TedTalk, "Why Our Screens Make Us Less Happy." One pronounced factor he noticed is how much of our personal time is being taken up by screens. Each year since the iPhone was introduced, more of our time is being spent in front of screens, with less time to pursue other enriching experiences, such as hobbies, time for reflection, and relationships with those around them.

The enriching experiences online are impressive, but most sites that people use for extended lengths do not produce beneficial behavior that improves mental health. People feel good about visiting, relaxation, exercise, weather, reading, educating, and health sites. However, many individuals spending time on dating, social networking, gaming, entertainment sites, and web browsing are not finding the practice fulfilling.

According to Alter, one of the reasons we are spending more time in front of our screens is because we no longer have cues that let us know it is time to stop—the end of the newspaper, episode, chapter. Now, the feed, the show, the texts just keep going. We need to create our own stopping cues. He has decided to never use his phone at the table. Other stopping cues include not using technology for the first hour of the day or putting your phone in airplane mode during the weekend. He found that those who started using stopping cues felt much happier and started incorporating even more into their lives.[18]

Without taking steps to limit our screen time, this loss of joy becomes a horrible cycle because, in our unhappiness, we turn to technology to make us feel better, giving up our privacy and security as we share more and more about ourselves online. We need to realize that ultimately our technological devices cannot bring us true joy. We need to connect back to our true selves and remember what really brings happiness and meaning into our lives.

Loss of Society

Technology is not only negatively affecting our personal health and relationships, it is also negatively affecting our society and democracy. As we spend more and more time on social media, we are more easily influenced by the content on our screens.

Much of this content isn't even created by real people. Artificial intelligence can create fake content and disperse it on social media. "Machine learning systems are already proving spookily capable at creating what are being called "deepfakes"—photos and videos that realistically replace one person's face with another."[19] This content is hugely influential and intended to sway users with targeted messages. Often, they have political aims and undermine our democracy.

Social media giant Facebook is a powerful tool for those who want to influence society's philosophies and individual's actions. Beyond that, the platform and its data have been used in malicious and illegal ways that undermine the democratic process. "Facebook, already facing deep questions over the use of its platform by those seeking to spread Russian propaganda and fake news, is facing a renewed backlash after the news about Cambridge Analytica."[20] Political consulting firm Cambridge Analytica has been accused of violating U.S. election laws after it accessed personal information on 50 million Facebook users in the U.S. without their consent. ABC news acquired a video of Alexander Nix, one of the founders of Cambridge Analytica, stating that "we were able to use data to identify that there were very large quantities of persuadable voters there that could be influenced to vote for the Trump campaign." Christopher Wally, an early employee of CA, told ABC News that the company planned to amass mountains of information on Americans. "We would ask people to fill out psychological surveys, that app would then harvest their data from Facebook, then that app would crawl through their friend network and pull all of the data from their friends also."[21] It really is impossible to keep your information private on social media, and you're not just putting yourself in danger when you use sites like Facebook.

During his View From The Top talk with the Stanford Graduate School of Business, Chamath Palihapitiya, founder and CEO of Social Capital, claimed that social media businesses "have created tools that are ripping apart the social fabric of how society works...the short-term, dopamine-driven feedback loops that we have created are destroying how society works."[22] In our rush to get our next dopamine hit through new Facebook likes and Twitter re-tweets, we are losing our ability to communicate and connect in authentic, meaningful ways with one another. Rather than truly developing our relationships with others, we are collecting followers and focusing more on our Internet image than on our true relationships and true selves. As we let our dependency on technology grow, we are losing more and more of ourselves.

Do you wait until it hits home or do you take steps to secure your life and identity now? We are creating our worst nightmare. Do you want to reclaim your life or do you want your content data public for anyone to do with as they please?

Develop Your Unhackable Mindset!

Now that you are starting to understand how technology can negatively impact your life and the lives of your loved ones, it is time to look more closely at how you can develop your Unhackable mindset today. Go to your *Unhackable Workbook* and complete the exercises and questions for Part II. Within the workbook, identify the greatest threats to you, your family, and your business. What are your vulnerabilities? Pinpoint specific sites, apps, and devices that are drawing you in, gathering your data, and putting you at risk.

To access your complimentary *Unhackable Workbook*, go to www.GeorgeMansour.com/workbook

Part III explores the powerful internal and external steps needed to reduce your dependency on technology and limit data gathering.

PART III

DEVELOPING THE UNHACKABLE
MINDSET

How to Secure Yourself, Your Family, Your Business, and Your
Community

7

TRAIN YOUR BRAIN TO BE THE SECURITY SHIELD

"All of a sudden you realize that you are the person who has control of your life."
—Jim Henson

I t is imperative we restore the original vision of an information-sharing highway that allows us to share freely, instead of hindering our advancement. An Internet paved with honest intention, just as the original inventors had planned.

These same devices are becoming impressive decision makers, with vast amounts of big data stored for future uses. Forget about the proposed advent of artificial intelligence and think about the ramifications of what's happening now.

The paradigm shift will repair privacy in our life. You can learn to control technology, not have technology control you. It's time to bring forth true security in an unsecured world. Are you ready to develop the Unhackable mindset? Will you stand in front of the wall we build together and reclaim your digital freedom?

Civilization is producing machines that try to think and feel for us, and we call it smart technology. Then we are amazed by their power and ability to accomplish menial tasks. Think back for a moment, when we had address books, they held the names and numbers of people you contacted regularly. But how many times did you actually have to look up a number? Most of the time you knew upwards of 50 to 100 numbers. Can you do the same thing today?

Humans seem to be amazed by the advancement of technology and subscribe abilities that our tools don't possess because we have misunderstood several fundamental points. People, in general, are not inarticulate or slow, nor headed for the evolutionary junkyard. Computer productivity is an inadequate model for an analogy to understanding the human brain. Eliminating the human equation to assure more speed, accuracy, profit, or protection is a sorry model for the future of society.

No hardware, no software, nothing that can ever be invented will truly secure us. The only way to true security is through yourself—the user.

The reverse engineering process must start from the bottom up, one step at a time. When an architect designs a blueprint, each plan has specific instructions for building the structure. Contractors don't build the roof and install windows immediately. Instead, they pour the foundation and construct the building from the ground up. Internet security works in the same fashion. Once the information is safely contained and eradicated, the user can begin to restore, reconstruct, and rebuild a solid foundation using proper connection protocols, avoiding all of the traps of the digital revolution age.

Personal Transformation

We are looking everywhere for answers, but we don't see that we are the solution. Stop being afraid and trust in yourself. Users do not have enough time. We think the machines will increase our productivity, but in reality, we're creating new problems for ourselves by not facing what we need to do right now. Modern devices are effective for one main reason: they're not burdened by human emotions. Machines cannot reason instinctively or feel bias or belief in particular sensations. Our intellect allows us to convey goals and pursue life. Users must stay in the driver's seat and control the outcome. We can add it to our toolset but never let it run without our interactions. Don't give away your private key to the front door. We need to neutralize all threats today and become a generation of neutralization.

Ask yourself a question: do the devices pursue human interactions? Do we believe these things to be unachievable, unknowable, or worthless? If not, when are we going to shift our focus?

We need to start asking, what is a successful collaboration between humans and machines? Can you imagine what that would look like?

Any continuous action becomes an instinctive habit. But at some point, we must enact our human connection, stopping the unconscious behavior that is invading our minds and society. We must become the practitioner of our problems and begin mending each facet of the inter-connected life. It is the equivalent to becoming your own virtual assistant. Only then can we create a happy and healthy existence with technology.

Theory and logic alone are not viable without establishing a fluid, dynamic environment. The belief and the environment together offer you a systematic security program. You need a fluid mindset that moves or changes easily depending on your environment.

Users must learn to work in a fluid movement, with the ability to change at any second. True security is only available through the user's hands. We are the interface protection layer which is the physical vessel and captain of our own boats, as we steer and navigate all unknown currents. The guide presented in the following chapters helps to bring balance and intelligence about Internet connections from data to life. The Unhackable mindset will help filter cravings for technology and the yearnings of social media connections.

Humans are sensitive to change, so we stay together in a central area, lying prey to the hackers waiting silently to attack. You have to break away, which is the change itself. Unfortunately, fish stay in groups to protect themselves, but we have to separate so that we gain our safety first by taking a step back into the future. Fortunately, it isn't about big changes; little changes can make just as much of a difference. Change is possible, but true transformation can only start with you.

In order for users to develop an Unhackable mindset, change is a necessity. Once the altered state of mind takes place, clarity with a defensive approach is obtainable. First, you need to learn the steps required to counteract the hacker's scams.

A Time for Change

Peter Singer, a Defense Department adviser on space threats and author of *Ghost Fleet: A Novel of the Next World War*, shares with CNN in *War in Space: The Next Battlefield*, "what any defense planner will tell you is, don't look for the ideal outcome, plan for the worst day so that you can survive it."[1] We need to be prepared for what the future might bring. With technology, change is inevitable.

The Internet brought about major changes to societies all over the planet. We went from clusters spread over the Earth's surface to a large aggregation interacting in a fashion never seen in civilization. It allowed people to connect socially without leaving the confines of their homes or location. Yet, we were able to maintain our independence. In seconds, distance meant nothing.

As we separated into different factions online, tightly

organized groups formed, moving at lightning speeds to capitalize on the opportunity. The exciting prospect forced a movement that developed multi-level conglomerates who closely schooled the masses on how collecting information in one structured fashion could benefit everyone. Before long, the entangled algorithms of the World Wide Web forced a massive crusade against the individual user to either conform or be left alone, exposed to the dangers of cyberspace. The unsolicited invader brought anxiety to a new level when they threatened the user by withholding personal information as ransom. Either the person pays the fee, or their private data is exposed to a structured environment to be used as they see fit. The insidious clutches of the hackers will go to any lengths necessary for the collection and solicitation of our precious digital information.

Society must change the destructive tide seizing our individuality. It's time to win back our independence, reopening the free information highway for everyone to enjoy without the demand for data collection.

As Chamath Palihapitiya advised, "if you feed the beast, that beast will destroy you. If you push back, we have a chance to control it and rein it in. It is a point in time where people need a hard break from some of these tools and the things that you rely on."[2]

Defeat Fear

Some users are paralyzed with fear when using the Internet and completely avoid new technology. Yet, others continue to move forward without hesitation. Some individuals even connect their children's toys to the Internet, leaving them exposed to the reckless behavior roaming online.

It is time to take a stand and deal with security issues, face the problems head on, and stop the cyberattacks. The resolution is very easy, but most users are complicating the situation.

Remember, it is a game of strategy, but true security starts with you.

Weakness, failure, and faults are there to strengthen us. Many great achievers say that in a lifetime of setbacks and comebacks, the greatest sense of accomplishment is not found in the realizing of your goal, but rather in the will to get up and continue when failure breeds doubt.

Imagine and embrace the concept of developing your Unhackable mindset today. The change starts with your thoughts. When your ideas are positive, it becomes easier to overcome the struggles and dangers you face online.

Be in the Moment

When behind the wheel of a car, we must be aware of our surroundings. The same applies to your network, computer, or phone. Slow down and notice what is going on—that will be your first secure move. Our drive for instant gratification causes many of the problems users face with online Internet usage. We try to out-speed the computer and click rapidly, impatiently wanting to see the display. In the process, we end up on pages that could be dangerous.

Let's stop for one second and be conscious of our actions. An active thought process will allow us to move forward safely. When anxiety takes over, it clouds all cognizant reasoning. We become blind to reality.

Technology has forced an urgency in our daily activities, causing inconsistent behavioral habits. What do you think would happen if we tried to finish one thing correctly? When you have an issue and accidentally download something, what is happening? What did you just do or click on? What can you do to avoid this problem in the future?

The current mindset is usually finding the easiest option and moving on. Technological advances have forced us to react in fear over a problem. Instead, we need to come from a point of strength and knowledge.

Be Consistent & In Control

Repetition teaches us consistency and is essential while learning a new subject. Just as our children need consistency, practice, and structure, we need the same thing in our digital life. You have to be unwavering in data control.

Often, we allow technology to make decisions for us. Most toolbars and pop-up software interfere with an Internet search or your browsing activity. It wants to direct all traffic through its own news outlet and starting page by adding a filtered connection to your searches, but the interaction slows the progress, keeping the user engaged and away from a natural, unfiltered connection. The browser is trying to think for the user. This is wrong.

Does anyone ask why every website and company needs you to create an account? They all require information to access your data. We are being blinded by the numbers and false hype. It's not about the user anymore, but the control over data collection.

In the event of a computer problem where the situation requires help from a tech expert, be consistent. Locate one person or advisor who can control and navigate everything. Go to a trusted business with a solid personnel department. Multiple techs and phone calls can morph a problem into a worse situation. So, you get a temporary solution and the real issue goes unchecked. If you had controlled the situation, it would be done correctly, and the environment would be preserved and intact with the least changes possible.

The Twelve Checkpoints

Security starts and ends with the user. We think hardware, software, and now built-in behavior-based security learning will do it for us. Time has proven that this is wrong. Look at all the high-cost security systems being used by the powers that be, yet nothing is secure.

We have to be more aware of the choices made with technology. It is the only way to monitor, test, and explore what is happening. Nobody takes one second to stop, think, and connect. If we did that, we could take the power away from the hackers. Learn how to recognize a scam and the hacker's ability to expose their victims. Then you will not operate from the fear factor or any indecision but from the strength of your own natural given talents as you acquire all the correct skills.

Internet connection is a manipulative and powerful force with the capacity to affect the character development and behavior of its users. The effects can be seen whether users are in a public, private, or a professional setting.

The World Wide Web plays an extensive part of our existence in society. Traditional business has been bypassed with the invention of the Internet infrastructure. People from around the world can communicate to conduct professional services anywhere on the planet. It is without a doubt the greatest invention of our time. But it has turned humans into social media slaves who surf the super information highway, void of any thought of the consequences.

The following section contains twelve short-term checkpoints designed to restore the original programming while filtering malicious intent and unintended negative consequences forced on users for profit making-purposes.

Once the user establishes intuitive patterns, they can learn to maintain self-enforced discipline and keep away from the indulgences of Internet life.

1. **Helplessness**—The admission that technology has enslaved me, and I'm dependent on a data-driven lifestyle. Could I turn off my data-driven life for 48 hours?

Technology has more control over us than we realize. We have become adrift in the sea of informational clutter, distracting and carrying us away from our neutral center. We are no longer captains of our technological autonomy or in command of our short-term realities.

Challenge: Start acknowledging this dependency in your personal and corporate life.

2. **Connectivity**—The understanding that we are all connected; my actions impact others. The realization that my dependence on the Internet affects me and everyone around me. Do I verify the validity of an email before it's opened?

The digital age has connected all of us. However, these connections are not being used to lift us up. We feel so compelled to stay up to date with new apps and social media that we are hurting each other. We enable one another by thinking this is what we want, when we feel exhausted and are craving change. In order to help one another, we must get off our current path and reconstruct our relationship with technology.

Challenge: Seek harmony in getting everyone to embrace this strategy as you involve others in this technological transformation.

3. **Commitment**—The commitment to creating a quality of life that is in my control and not dependent on technology. Am I able to commune without technology and be mindful at all moments?

We must commit to finding our center point and get back to simpler times. Only we can captain our ships to navigate through the murky waters that technology has left us in. To start, we must lay out a guide to follow in order to stay on track while measuring our progress. In turn, we start to adjust our short-term reality so that we may rebuild a new relationship with technology long-term.

Challenge: Set up a plan to stay on track and commit to being mindful about technology use.

4. **Introspection**—The examination and observation of my processes to discover weaknesses and strengths related to

technology. Am I aware of my environment while connected to protect my privacy and security settings online?

Reflect on what aspects of technology are exploiting your weaknesses. We need to step back and understand how technology triggers our emotions and causes us to act in an almost predetermined way. We must understand the pitfalls we keep falling into in order to avoid them.

Challenge: You should have parameters in your plan for uncontrollable changes like supply cost, economy, and labor issues.

5. **Weakness**—The resolve to uncover Internet deficiencies and weaknesses to learn about common failures. Do I stop to read license agreements?

We must lay bare the deficiencies technology is exploiting from us. This can be difficult, but it is necessary so that we may discover solutions to fill the cracks in our foundation.

Challenge: Form practices that are related to oversight that verify benchmarks are being attained according to schedule and monitoring activities.

6. **Desire**—The desire to avoid trigger points that generate the cravings and yearnings of my digital life. Can I stop clicking on those pop-up ads, and think before I act?

We are familiar with the basic concept that life is hard, yet accepting such a simple principle can be a challenge itself.

Challenge: Develop updates and revisions to stay vigilant and build periodic reviews of the plan. Put into practice plans for goals, benchmarks, and monitoring.

7. **Seeking**—The seeking of knowledge on how my digital life has been influenced by my daily routines. Neutralize and change these points to work for me and not against me. Can I control the technology or does the technology control me?

We have established an overarching problem, but now we must pin down the specific mistakes being made that are applicable to us. Starting with small, common-sense solutions can be the best way to get the ball rolling.

Challenge: Long-term goals must never be left out as they develop the perimeter of all security.

8. **Willingness**—The willingness to learn and grow while keeping my sensitive data secure. Am I ready to make the necessary changes and stay away from the questionable things online?

As we have become so entrenched in our bond with technology, severing certain connections will feel monumental. If we desire true change, we must be willing to sacrifice our daily conveniences in exchange for the clarity and security we have lost along the way. There is no such thing as a "free lunch." The privacy and freedom we yearn for can only be taken only if we reduce our dependency on the easy access we have become accustomed to.

Challenge: This strategy ought to be seen as a daily action instead of treated as an isolated plan.

9. **Life**—The implementation of new modifications in my daily life will directly affect my digital footprint in a positive manner. How can I modify my online behavior?

If you maintain these new changes and complete the guide you have created, then the simpler life you sought after will start to follow.

Challenge: The finest plans must not be overpowering. They need to be achievable and actionable.

10. **Continuity**—The maintenance of ongoing, focused patterns of behaviors that correct my actions in daily life. The desire to always be in full control and sustain power over my decisions to the connected world. Am I ready to stop the tracking of my sensitive data?

You created your own guide, thus becoming an agent of change to combat the tide of technological bombardment. The grasp technology has over your life diminishes and you begin to rebuild your insulating shell to keep it from taking over. The restoration of your center point is no longer a concept, but a reality you have made possible.

Challenge: Advancement reports and the reaching of benchmarks need to be noticed continuously.

11. **Contemplation**—The commitment to anticipate how I consciously interact over the Internet. To always be aware of my habits when dealing with new technology. Can I dedicate the time to make the necessary changes, and regularly change my passwords?

By taking the steps to reconstruct your relationship with technology, you have created your own new center point. From there, you can rebuild your security and privacy and be selective about what information you want made public. You are no longer adrift; you have found your port to drop anchor. Having this foundation means you can move beyond the guidance and start consciously being able to anticipate new obstacles technology may present.

Challenge: Entities need to have enough authority and means to implement the blueprint plans.

12. **Collaboration**—The responsibility to become agents of change to the things I have learned. Collaborate in a unified fashion and take back control of our sensitive data and online identities. Technology should allow us to connect, share, and grow, as it helps our life become less complicated. Am I ready to manage my own data and keep my personal information private?

Regaining your clarity and security will bring you confidence to move forward and not fall victim to bad actors who would lead you back into the rocky waters. You understand what is necessary to maintain your privacy and peace of mind from the draining technological wheel you were stuck on. More importantly, now you can act as a model for others to follow. You can display how to access technology as a tool to move the world forward. You have reclaimed your freedom and established the Unhackable mindset to maintain a conscious connection with technology and be the master of your own destiny.

Challenge: Use these strategies daily as you bring it all

together. Stop looking over your shoulder and allow it to be your code of ethics in the here and now.

Restore by educating yourself! Users must free themselves from all online intelligence, data gathering information, controls, filters, cookies, etc. The action will bring forth a paradigm shift designed to survive the test of time and speed of innovation. It raises us beyond the failures of today's complex technology.

Reconstruct your private offline identity. Begin initial steps toward rebuilding and renovating life online for all IoT devices, etc. Create transformational mindfulness throughout the Internet, in a reinvented world full of incisive decisions. The user can learn to control technology, not have technology control the user.

Rebuild your life online by looking through a different lens. By establishing an interconnected shell, the user can build a conceptual demarcation line that creates a segregated stone wall defensive system. The masonry is mortared by the individual constructing the wall that will help stop threats to the user's sensitive data. It's time to bring forth true privacy and security in an unsecured world. Are you ready for the Unhackable mindset?

By acquiring the skills to outmaneuver the processes of today, as well as future electronic database expansion, we can reclaim our digital freedom, thereby defeating communication technology with great ingenuity.

The Unhackable mindset creates the conceptual model to enact a multilayer digital union between technology and human interaction. The insecurities and impurities that threaten us online with every new connection must be controlled through a collaborative approach that will suppress the threats to our sensitive data.

The Unhackable mindset aids users in breaking the connective patterns of Internet programming that will help build a stone wall of protection working in your defense. To succeed in today's world, one must adapt to and accept the evolution of technology and the responsibilities that come with it. We can become the captain of our vessel. Do you want to be in control of your smart data?

Progress requires patience. *Kung Fu* is a Chinese term referring to any study, learning, or practice that requires patience, energy, and time to complete. This is our paradigm shift—through hard work, patience, and time, we can restore privacy in our lives.

8

STOP TAKING SHORTCUTS

*"If you are willing to do only what's easy, life will be hard.
But if you are willing to do what's hard, life will be easy."*
—T. Harv Eker

On a recent vacation with the family, my kids and I
started getting cold while hanging out on the first
floor of the hotel. I had the stroller with the baby, so
we decided to take a shortcut to the second floor instead of going
through a path that took us outside in the cold to get to our
room. I wanted to reach the second floor quickly. I said, "Kids,
let's go up from here. I think there's a shortcut this way."

We kept going around in circles, but there was no way into
the other section. It was like a labyrinth. There was one way in,
one way out. The only way to the next floor was to walk outside
and in the front door of the other section. We ended up starting
back at the beginning.

It was a great lesson for the kids and a good reminder for me.
People want shortcuts or the easy way out, but that only leads to
more work and time wasted. We must do what's hard before it

gets easy. I wanted a shortcut to be warmer, but instead, my decision cost us valuable time. We had to go through the cold anyway, and there was no shortcut.

Just like the situation with my sons, we want quick, easy security solutions, but at the end of the day, we are actually creating more problems and not making things easier. Instead, we fall quietly into the abyss with all of this technology, lost with no way out. Will we connect back to the real meaning of technology, invention, and connection or will we lose control over our lives, minds, and privacy?

Be Informed and Work with Professionals

My big focus is empowering and teaching my clients how to protect themselves and make informed decisions in times of distress. If you do what's hard in the beginning, then you will have it easy later, but if you do what's easy in the beginning, you will have it hard later. Don't allow a lack of time or not enough resources to compromise you. Never be in a rush or afraid to ask for help, because if you set up your environment correctly and have a backup system, that will take the element of surprise and panic away. No one wants to scramble while trying to fix a problem. Be in the driver's seat and never operate out of frustration, or the situation will control you.

It's imperative to ask yourself these questions every time there is a problem:

- What is my current issue?
- What is affected?
- Has this happened before?
- When was the last time it happened?
- Has there been any changes since?
- How can I prevent this from happening again?

The Unhackable mindset is going to teach the user how to

neutralize their Internet environment. Knowledge is power. As you learn to act and move, you gain more wisdom. As we sacrifice our short-term controls, like our decisions and choices in order to feed our quick gains and instant gratification, thinking to gain more time and speed, at what cost are we doing this? We are only disguising and prolonging the inevitable problems we all need to face. Do the hard parts first; step back into the future. We're building our foundation on quicksand instead of making ourselves the cornerstone that holds it all together.

My strategy is always to get to the root cause, while also providing short-term tactics to plan, design, and deploy everything correctly from the beginning to avoid unnecessary headaches and cost. Without adequate planning, you will pay double for intervention later and must work overtime to catch up.

A common action is either to shortcut a problem or find a cheap or free solution. Sometimes, we try Googling the issue without proper understanding or direction. One of my clients, Eric, lost Internet access; his first action was to call the Internet carrier. In turn, they sent a technician onsite since the issue could not be resolved over the phone. As it was his first trip to this site location and without any prior knowledge of the network environment, the guy reset all the boxes like Wi-Fi configurations and network access and confirmed the Internet was working properly. He departed for his next job without testing or confirming anything else.

Eric was prepped and returned to work, but he realized the whole network had been reconfigured to default settings, none of the devices were interconnected anymore, and he couldn't print or share files. His rash judgment caused considerable, immediate emergency issues. In panic mode, Eric called me, frantic over the situation. Instead of reacting to a particular situation, we must follow a mental checklist to determine all potential issues, then decide on the best course of action to resolve the problem before jumping into a rash decision that only creates a problem on top of another.

We cannot solely rely on the mechanics of hardware and software of cybersecurity; we have to focus on the psychology behind our decisions. Rather than utilizing a quick mechanical fix, take the time to truly understand what is going on and get into a preventative mindset.

The 80/20 rule has made appearances in the business world for over 100 years. The Unhackable 80/20 mindset rule is that 80% of the work you do to stay secure is psychological and 20% is mechanical. Once you have taken the time to master a proactive mindset, you will rely less on mechanics and make more informed decisions about how to best use external tools in preventative ways.

Sometimes, users will pay upfront to have a problem fixed correctly, granting them peace of mind later. If you are in this situation, using a company who has a solid reputation or serviced your computers in the past is an ideal setting. I know when the problem is explained it will be addressed without complications. Clients' computer systems are personal and private, and they must be treated with respect—a service I take very seriously. The action has created a consistent environment where the alterations are minor, which allows me to guide them correctly.

One of the first protocols is to call the tech advisor while they are stuck until they can achieve the Unhackable mindset. It is the only way to correctly interpret the situation. However, many decide to try and fix the problem quickly and only make matters worse. In their attempt to save money or cut corners, the problems only escalate and cost additional money and time. Once you have completed the Unhackable course-correcting process and plan to understand the mindset and handle the situation, making correct decisions is not an issue or fear factor anymore.

What trick or scam did I fall for? Most don't ask this question.

We all need to work toward a common goal and understand that it's not just about fixing the current problem, but identifying the root causes and knowing what we need to do to avoid them in the future.

Society has fallen into an instant gratification mindset, but at what cost? Human behavior has been altered over the Internet; everything is out of order. Users must educate themselves prior to connecting. It's the only way to find real balance in the cybersecurity era.

Shortcuts Remove Understanding

Do you want control of your digital identity? If the answer is yes, then stop looking for shortcuts and build a foundation of knowledge and understanding before, during, and after any connection. This will restore your natural, true power back in your hands. Some people cannot function without their electronic devices for one second, let alone an entire day.

What kind of message are we giving our children? Do you want the adolescence of society growing up without the capacity to think decisively? Our fixation with technology is creating a generation of non-thinkers. The subject is a passion of mine because I have three children. I want them to have decisions and choices, and not live in a world controlled by machines. Let's neutralize all threats together in a joint collaborative effort to secure our individual borders.

The problem does not stop with individuals, but carries forward to businesses. When was the last time you went into a store and the clerk knew how to count change? Unless the cash register tells them the exact amount, they are lost. Society is losing control right before our eyes, and we remain blind to all the incognito actions of governing factions.

One particular example comes to mind: I was at Dunkin' Donuts, and a customer wanted to pay through the app, but his app wasn't working. He ended up having to leave without any donuts because he did not have any money with him. What happens when we lose power, and the Internet goes down? Do we expect it to always be up and running in a time of insecurity? We want to e-pay everything, where it stores our personal information and content data. This may seem like a great shortcut, but what happens when it doesn't work?

When technology stops working, we don't know how to do anything manually anymore. People have become complacent in many ways. The current generation has everything easy; they don't know what it feels like when you accomplish a great task.

True Change

Do you want to secure your digital life? Everything is about instant access, but do we truly need constant updates about our social media page, or can we live with reports a few times a day? True change comes when a person is ready to step forward and accept the need for an alteration in their daily habitual patterns. Our human behavior becomes mundane. People don't like change and are usually attached to past experiences and behaviors. However, our technological environment is ever-evolving. When emotions are a factor, tensions increase, and people stop thinking rationally. They just want everything to be done right now, without delay, seeking no other alternatives. How do we stop this rat race?

Most of our conflicts are not external, but internal. We are moving too fast and don't know how to slow down. You need to change how you look at cybersecurity. We can't just take the elevator to the top. The process must be completed in stages and steps, with a clear and vivid understanding of the lesson prior to moving forward.

Our thoughts are energy, and physically change our mood. Therefore, we must create an atmosphere of neutral emotions. The environment will breed positive change and balance.

• • •

Once the user has established the correct mindset, existing problems will become apparent. However, in circumstances where we start to feel anxiety over something not understood, it's time to step back, take a deep breath, and relax. If, at that point, the problem cannot be resolved, contact an advisor you trust who can handle the situation. We must think before we act, taking the time to learn the thought process before jumping into unfamiliar waters alone.

9

BALANCE THE OLD AND THE NEW

"Balance is not something you find, it's something you create."
—Jana Kingsford

Choose Security Over Innovation

As a society, we are focused on innovation, which can be infectious. Companies are moving quickly to deliver their product by flooding the market with new technology to make life easier, but are the advancements doing as the marketing suggests?

There is now a field of study dedicated to analyzing how technology persuades users and influences their attitudes and behaviors. Stanford Persuasive Tech Lab shares:

Captology is the study of computers as persuasive technologies. This includes the design, research, ethics and analysis of interactive computing products (computers, mobile phones, websites, wireless technologies, mobile applications, video games, etc.) created for the purpose of changing people's attitudes or behaviors. BJ Fogg derived the term *captology* in

1996 from an acronym: Computers As Persuasive Technologies = CAPT.[1]

As disturbing as it is to consider how influenced we are by technology, the Stanford Persuasive Tech Lab shares the optimistic reasons why it created this area of study:

> **Machines Designed to Change Humans.** Yes, this can be a scary topic: machines designed to influence human beliefs and behaviors. But there's good news. We believe that much like human persuaders, persuasive technologies can bring about positive changes in many domains, including health, business, safety, and education.
>
> We also believe that new advances in technology can help promote world peace in 30 years. With such positive ends in mind, we are creating a body of expertise in the design, theory, and analysis of persuasive technologies, an area called captology.[2]

Society is offering and adopting a new baseline of impulsive decision makers as we unburden ourselves. We put all our actions and trust into a device that is not alive and is now leading warfare, manufacturing, finance, and many other fields. Users should be opting for cautionary actions to produce changes that will help advance civilization instead of producing mindless actions from people that have lost touch with reality. We outsource all our responsibilities and jobs to a virtual chamber of code as it stores vast amounts of data for future use, control, and foreign intelligence. True advancement is not to offshoot or allow the technology we create more control, but instead it should improve the human condition. In *The Guardian*'s "What Does It Mean to be Human in the Age of Technology?," Tom Catfield explores the collective experiences we have online:

When I think about the future of human-machine interactions, two entwined anxieties come to mind.

First, there is the tension between individual and collective existence. Technology connects us to each other as never before, and in doing so makes explicit the degree to which we are defined and anticipated by others: the ways in which our ideas and identities do not simply belong to us, but are part of a larger human ebb and flow.[3]

When we tap into this collective existence, we are no longer making independent choices. Instead, every decision is influenced by the group, rather than the individual.

Common Cognitive Biases

Below are examples of some common cognitive distortions seen in depressed and anxious individuals. People may be taught how to identify and alter these distortions as part of cognitive behavioral therapy.

- Selective abstraction–drawing conclusions on the basis of just one of many elements of a situation.
- Personalization–"this is my fault," attributing personal responsibility for events which aren't under a person's control.
- Magnification–"Making a mountain out of a molehill"–blowing things out of proportion.
- Minimization–Useless, inadequate, failure. Downplaying the importance of a positive thought, emotion or event.
- Arbitrary inference–Drawing conclusions when there is little or no evidence.
- Overgeneralization–Making sweeping conclusions based on a single event.[4]

Innovation should be treated as an immature form as it transforms to a more mature state. If the user has enacted the correct thought process it will create true security awareness. Thereby, security defenses will help to build an active barrier to safeguard against malicious attacks. Innovation can only benefit humans if we learn about technology. Otherwise, the same machines we created to advance mankind will overtake civilization. Machines will only be as good as the programmer and code that created them. According to Catfield, "Our machines aren't minds yet, but they are taking on more and more of the attributes we used to think of as uniquely human: reason, action, reaction, language, logic, adaptation, learning. Rightly, fearfully, falteringly, we are beginning to ask what transforming consequences this latest extension and usurpation will bring."

The Unhackable course-correcting game plan outfits each user with the proper tasks that help them to focus on their technologically-challenged mindset. In return, it closes the gap between humans and machines; forming an improved set of communication skills.

Humans are living in a world that is almost totally connected in every way possible, yet most people are overwhelmed with the technology. Unfortunately, we keep moving ahead blindly without asking the right questions:

- What should I be doing with all of this technology?
- How can I use it securely?
- Am I safe?
- How do I use it and still enjoy the Internet?
- What should I be doing for my home and business?
- How can I make this easy on myself?
- How do I protect my family and work online?

Back to Basics

The proper use of technology requires neutralizing biases and influences as we gain balance to restore the relationship between human and machine. I am not suggesting we eliminate technology (it's a necessity in modern civilization), but we need to use it with established safeguard protocols. Just as children learn to crawl, walk, then run, the same applies to Internet users.

Safety should always be first. Identify your priorities and needs, and then adjust your interactions with technology to match your values. What is your time worth to you? What is the cost of the technological investment you have made? Do you feel you are being rewarded for this investment? I think most would disagree. That is why we need to make a change and get back to the simpler times we all crave.

You need to Restore, Reconstruct and Rebuild your real-time, encryption safe-mode that has been stripped away by a malfunction in the instructional coding. These three steps will become our shared community of practices.

Restore: We must restore the way we view and interact with technology. The current relationship is not beneficial for us; therefore, we must tear down what we believe works and seek a better path. This will become your short-term, crisis-driven pathway that develops through the crisis where you find your way in many different reorganized forms and mannerisms.

Reconstruct: We need to reconstruct our relationship with the truth in a positive manner that doesn't clutter our lives with meaningless information and interactions. Get off the hamster wheel of keeping up with the latest trends. Take a step back and seriously ask, *what do I truly need in my life?* This will become your short-term, crisis-oriented awareness where you uncloak all the misconceptions our culture has used to cover the original design.

Rebuild: We need to get back to the center point and the authentic life, where we have control of technological use.

People didn't always have these technological marvels and were able to live their lives more carefree, not weighed down by the suffocating pressure to stay connected. This is about finding the sweet spot where technology can enhance our lives without being too overwhelming to handle. I think this is ultimately what people want. Transform your technology as you use the current crisis to draw you into a very focused, intentional, long-term, helping relationship, where you journey a lifetime together with technology as a powerful tool, hand-in-hand.

Reconnect to the Past

What happens when technology breaks and everything we have built fails? We will be stuck.

In a grid down situation, society will be at the mercy of broken technology. The innovation will only continue to overtake civilization if humans do not pause the control over our lives. We will have no way out if our past has been erased. Therefore, we must merge the past with the present to bring unity back to civilization. I believe that we're nurtured to behave in certain ways, sovereign from our genetics and natural (nature) starting points that determine our baseline behavior, personality traits, and abilities.

As generations pass, small amounts of human history are being washed away. This has forced us to move forward without truly understanding what came before us. The lack of knowledge will eventually boomerang, forcing chaos in society. We are already seeing massive challenges being forced upon civilization by a lack of true cybersecurity.

Our digital identities give us a true challenge, but instead of focusing on the negative, let's use this as fuel for a true awakening, to figure out what place we occupy in the Universe.

The Unhackable mindset will create your new systematic plan, through quality assurance protocols and methods that will prevent mistakes and defects. It teaches the correct usage of technology with a set of clearly defined steps. Now is the time to restore the original program away from all malfunctions. Society cannot push forward and secure our private data unless we have obtained suitable education and training awareness in cybersecurity that will create a true alliance between humans and machines.

10

REINVENTING BUSINESS TECHNOLOGY
& SECURITY

"Technology does not run an enterprise, relationships do."
—Patricia Fripp

Cybersecurity has played a key role in the last three decades. Private and public sectors of business institutions have spent an endless amount on trying to secure private information. Simultaneously, authorities are creating laws to institute regulations that protect our precious data. But none of these actions can truly guarantee your identity is protected. A solid defense shield can only be achieved when the user takes full responsibility to be proactive with their behavioral thoughts and actions. Big data is continuing to grow at a rapid pace. Our digital footprint is at serious risk, due to constant connection.

TechRepublic's Jason Hiner's article "Is Perimeter Security Dead and Is Protecting the Data All That Matters?," reviews the security measures corporations have taken in the past and makes the case for approaching security from a different perspective:

The traditional security model in IT has been to build a big wall around the corporate network to keep intruders out and to let the people on the inside have privileged access. That model is breaking down because networks are losing their borders.

The primary method of corporate computer security over the past three decades has been focused around the network. It's been about allowing those inside the network to have privileged access to corporate resources and building impenetrable walls to keep outsiders out. Unfortunately, this model is rapidly losing its effectiveness because the borders of networks are becoming much more fluid and dynamic with the advent of VPN, Webmail, push e-mail on smartphones, telecommuters, and a geographically dispersed and mobile workforce.[1]

Everyone just wants to get their work done and they don't take time to consider their privacy and security. Are you like that? I always hear, "I have no time! The most important thing is to keep the computers up and running!" This kind of mentality sets businesses up for disaster. Businesses need to be smarter about interactions with technology. Security should be more important than convenience.

The Free Dictionary provides this definition and examples of influence: A power affecting a person, thing, or course of events, especially one that operates without any direct or apparent effort: *the pervasive influence that TV has on modern life; young people falling under the influence of a radical philosopher.*[2]

Neutralizing the influences over our data and how it flows online is very important in getting users ahead of all these sneaky phishing scams, which are only getting more sophisticated and destructive.

Many surveys show that a lot of companies are moving to the cloud faster than ever. Executives have differing opinions on whether to adopt emergent technology early.[3]

Advancements in mobility, social media, and cloud computing are changing the way we work, interact, and make purchases today. This is all great for businesses as it brings growth and profits, but it also brings a lot of risks. Businesses are conducting billions of online transmissions and transactions daily. With so much data online, businesses expose themselves to an increasing amount of cyber theft, online attacks, and data breaches. The media focuses so much on data breaches of big corporations, but small and mid-sized businesses are also at risk. Cybercriminals assume smaller business owners don't have the resources or expertise and deep pockets that big companies do to defend themselves.

With any breach, you can lose your information, expose company secrets, disclose customer data, and reveal confidential employee information. Most companies rely on technology to do everything. Failure to teach personal security leaves the company very exposed to attacks, which leads to the loss of valuable intellectual property, reputation, revenue, and data. We need a transformation in all areas of business technology, with the workforce and security. Transformation matters now more than ever, but not where we think it needs to be.

Businesses focus on Return on Investment (ROI), but we forgot one thing: are we doing it for the money or to improve the human condition of the consumer? If we focus on serving the customer, then the money comes automatically. However, once we start to plan projections, we're getting in the way of our purpose and it becomes not about the consumer but about the product. When new products are released, companies have functions and features waiting to be unlocked at a future date. To what, improve their profit margins? This isn't truly serving the consumer. We need to stabilize the environment and minimize change as much as possible.

Many focuses are geared toward the ROI instead of our well-being. When profits run us, we move away from what we should truly be doing. So even though we have seen so much change, is it positive change, busy change, or negative change? If the environment has not been properly prepared, is there an immediate need for these constant changes?

We have this digital transformation happening where we modernize our infrastructure to reduce operational costs and overhead, and then we automate processes for delivering services. Then we transform our infrastructure operations to increase business speed. Our technology is moving so fast, and then we ask ourselves why we have so many security issues.

Over the past century, the average lifespan of a company listed in the S&P 500 index went from 67 years in the 1920s to just 15 years in the 2010s.[4] Why such a decline? We have mergers and acquisitions; companies have fallen and failed. Technology has changed, and so new markets and business models pop up and legacy companies just fade away. The business world has changed due to the speed at which technology is advancing and reshaping the world.

When you have a robust digital transformation strategy, it will help you overcome these challenges and be ahead of the flow. What is needed is employee training programs that will build awareness and change behavior to better improve your security shield. One uninformed employee can expose everything and bring the whole system down. Your business's security is only as strong as your weakest link—one uninformed employee puts the entire company and its customers at risk. You need to create an environment that fosters privacy and security above all things. Again, what do you choose, security or convenience?

Business Security

Users must learn how to secure business information in an ever-changing digital age. The environment is not equipped to handle the increasing growth of technological advances. Laws and regulations are being activated, but by the time each one becomes effective, the hackers are out of reach. Dangers lurk in every area of the Internet between public and private sectors, even the tech companies aren't able to create software, hardware, and next-generation firewall products in time to protect the users. Therefore, it's imperative to avoid the reactionary mode and create a proactive, safe environment.

The Unhackable mindset provides steps on how to overcome the default configurations before, during, and after any event and neutralize the situation to prevent potential problems rather than responding retroactively. You can't depend on regulations, security software, or hardware to protect your constant connections. Everyone in your company needs to have the knowledge to make informed choices and decisions about technology and its safety. ThinkHR's article "Cybersecurity: Employees Are the First Line of Defense" shares common cyberthreats aimed at employees:

> **Phishing emails.** These are relatively unfocused email messages designed to collect sensitive information, such as login credentials, credit card information, Social Security numbers, and other valuable data. Phishing emails pretend to come from trustworthy sources like banks, credit card companies, shippers, and other sources with which potential victims have established relationships. More sophisticated phishing attempts use corporate logos and other identifiers to fool potential victims into believing the emails are genuine.

> **Spearphishing emails.** These are targeted phishing attacks typically focused on one company or affinity group (such as an industry organization), reflecting the fact that a cybercriminal has studied the target and crafted a message designed to have a high degree of believability and a potentially high open rate.

> **Consumer file sync and share tools.** Productivity tools like Dropbox, Microsoft OneDrive, and Google Drive, which let users make files available on all desktop, laptop, and mobile platforms, are generally safe but can be targeted by sophisticated criminals as an entry point. For example, when an

employee accesses corporate files on a home computer that doesn't have current anti-virus software, the employee can inadvertently infect these files with malware. When files are synced back to the employee's work computer, malware can infect the network because it may have bypassed corporate email, web gateway, and other defenses.

Watering holes. In these social engineering attacks, cybercriminals identify websites they would like to infiltrate and that employees might visit on a regular basis. They infect these sites with malware.

Malicious Internet advertising (malvertising). This is designed to distribute malware through advertising impressions on websites.

User errors. Users sometimes inadvertently install malware or compromised code on their computers. This can occur if they install ActiveX controls, download a code, install various applications intended to address some perceived need (such as a capability that IT does not support), or respond to scareware attempts that prey on users who are trying to protect their platforms from viruses and other malware.

Mobile malware. The growing use of smartphones and tablets is increasingly being exploited by cybercriminals. Most infections impact Android devices.

Compromised search engine queries. Valid queries can be hijacked by cybercriminals to distribute malware when employees perform web searches. This type of attack relies on poisoning results, leading to the display of malware-laden sites during these searches. This is particularly effective for popular search terms, such as information on celebrities, airline crashes, natural disasters, and other "newsy" items.

Mobile copycat apps. Some mobile applications are distributed through vendor-based and third-party stores that offer varying levels of security. If the store lacks stringent standards, serious security risks like distribution of copycat apps and malware that can cause infections when downloaded can occur.

Botnets. These are the source of many successful hacking and phishing attacks against high-profile targets. A CenturyLink Threat Research Labs study for a 2018 threat report tracked an average of 195,000 threats per day from botnets impacting an average of 104 million unique targets, from large servers to handheld devices, that steal sensitive data and launch network attacks impacting businesses worldwide.

Ransomware. In this particularly malicious form of attack, a cybercriminal can encrypt all files on a hard disk and then demand ransom for access to a decryption key. Victims who choose not to pay the ransom quickly will have their files remain encrypted permanently. Cryptolocker, a common variant of ransomware, typically extorts a few hundred dollars per incident and normally is delivered through email with a PDF or .zip file disguised as a shipping invoice or some other business document.

Hacking. With this form of cyberattack, cybercriminals use many techniques to breach corporate defenses.[5]

Unless leaders invest in the human education of cybersecurity, hackers will continue to exploit weaknesses. It all starts with business owners placing shared responsibility, accountability, and transparency on employees for protecting key assets. Most employees are unaware of the threats, and others ignore security procedures in the name of productivity. IT security professionals are in short supply, and those who know what they are doing get in and out quickly.

The implementation of simple controls and security best practices can greatly improve the overall cybersecurity of a business. Online Trust Alliance reports that "93 percent of all breaches could have been avoided had simple steps been taken such as regularly updating software, blocking fake email messages using email authentication and training people to recognize phishing attacks."[6] What is needed are employee training programs that will build awareness and change behavior to better improve security.

Every member of a business should understand the basics of security. They should know what to do to keep the company safe and know how to react when they face a security concern. It is essential that every employee take greater ownership for their work and actions. They need to have a strong understanding of strengths and weak spots, which allows the company to provide the support needed to enhance their performance and security of their digital world.

Be Your Own Chief Digital Officer (CDO)

Professional hackers attack data information for many reasons: pleasure, personal, and business is just the beginning. It's all about bringing the correct people together in the right way to make real-time security, and I don't mean software or hardware. All organizations, groups, and users need to have more shared responsibility, accountability, and transparency for all data processing. Many high-profile cybersecurity breaches occur when hackers target an organization's weakest link: its people.

Business success should be measured by results. Nothing is secure. Why are we moving at this pace, while continuing to invest in hardware and software innovations, if we can't secure what's already in use? Society needs a new process and approach. Athletes are result driven, and we need to attack our data and information with that approach. We need to become managers of our own digital identity, be responsible for our decisions and choices. The user must have a clear expectation to overcome all these malicious attacks targeted toward our information and data.

It is important to understand how your business relates to the digital world. What started as initiatives led by CIOs and IT organizations have spread throughout every area of a business.

Executives in other critical functions are also leading investments to bring their own technology roadmaps to reality. There is always internal competition between groups. For example, CMOs (chief marketing officers) and CIOs (chief information officers) often compete in spending when it comes to new technologies and resources.

We first need to realize our system is broken because we are not working together. The departments that are attempting to move away in an independent fashion must stop and learn to collaborate as one unit with everyone else. Leaders must be the agents of change.

Businesses have CEOs, CMOs, and CIOs. Now, the newest corporate position being created is a CDO. The chief digital officer controls digital technology and marketing. They "are tasked with mastering the strategies and tactics to achieve and surpass your digital business objectives."[7] CDOs are predicted to take over and reach high numbers. "The race to drive competitive advantage and improved efficiency through better use of information assets is leading to a sharp rise in the number of chief data officers (CDOs). As a result, Gartner predicts that 90 percent of large companies will have a CDO role by the end of 2019."[8]

The CDO will be even more powerful than the CEO because everything is technology-based and digital. "Every budget is an IT budget," said Peter Sondergaard, senior vice president of research at Gartner. "Technology is embedded in every product."[9]

Some businesses aren't certain if they need to hire a CDO. No! The CDO is you and the employees underneath you. But it is the essential duty of everyone who comes in contact with the environment, no matter the confidentiality level their position carries. The collaborative effort falls upon everyone involved in the atmosphere, creating a new security blanket.

Instead of bringing in an outsider as a CDO, every member of the team should work together to identify the role technology will play in the business. Everyone needs to be aware of the business's goals and informed about how technology can be utilized to achieve those goals. Additionally, everyone needs to have knowledge about how to maintain this technology and use it safely. The article "Threat or Gold Mine? What Lies Ahead for Managed Services" presents questions that a business team should be asking its members as they create a solid tech foundation for the company:

> First, what combination and configuration of devices, apps, interconnections, and data will help them achieve a particular business goal?
>
> Once they know what they want to set up, how will those devices, apps, and data be set up, configured, and connected? Who will do it?
>
> And after a system is up and running, how will it be maintained? The technology of the future will break down, wear out, overload, overheat, throw errors, and just plain not work in the same way our current technology does. Somebody has to know how to fix it.[10]

Data Boss

Cloud computing is enabling business transformation as organizations accelerate time to market and business agility. New issues are being introduced as we move our data into a foreign space. Evolving cloud technologies and approaches, however, can create security gaps and human errors. Data protection rests with you and your organization as it flows together with the cloud provider. We need to become our own data boss. Most users believe that once their data is moved to the cloud, security and all hassles belong to the cloud provider. This is a false perception that has allowed users to become more vulnerable and created a multitude of new threats. The shared responsibility lies with both parties creating the correct ground rules for better accountability and transparency.

To begin with, your business should have clearly established security requirements, such as the example below from South Puget Sound Community College. Such requirements provide clear expectations and acknowledge the importance of educating and training all employees:

Computer System Security Requirements

This office shall maintain a security system covering its computers, including any wireless system, that, at a minimum, and to the extent technically feasible, shall have the following elements:

1. Secure user authentication protocols including:

control of user IDs and other identifiers;

a reasonably secure method of assigning and selecting passwords, or use of unique identifier technologies, such as biometrics or token devices;

control of data security passwords to ensure that such passwords are kept in a location and/or format that does not compromise the security of the data they protect;

restricting access to active users and active user accounts only; and

blocking access to user identification after multiple unsuccessful attempts to gain access or the limitation placed on access for the particular system;

2. Secure access control measures that:

restrict access to records and files containing personal information to those who need such information to perform their job duties; and

assign unique identifications plus passwords, which are not vendor supplied default passwords, to each person with computer access, that are reasonably designed to maintain the integrity of the security of the access controls;

3. Encryption of all transmitted records and files containing personal information that will travel across public networks, and encryption of all data containing personal information to be transmitted wirelessly.

4. Reasonable monitoring of systems, for unauthorized use of or access to personal information;

5. Encryption of all personal information stored on laptops or other portable devices;

6. For files containing personal information on a system that is connected to the Internet, there must be reasonably up-to-date firewall protection and operating system security patches, reasonably designed to maintain the integrity of the personal information.

7. Reasonably up-to-date versions of system security agent software which must include malware protection and reasonably up-to-date patches and virus definitions, or a version of such software that can still be supported with up-to-date patches and virus definitions, and is set to receive the most current security updates on a regular basis.

8. Education and training of employees on the proper use of the computer security system and the importance of personal information security. [11]

As your company works to mitigate cybersecurity incidents, make sure you have a protocol in place. TalaTek shares the following response steps informed by the National Institute of Standards and Technology's Computer Security Incident Handling Guide:

Incident Response Planning

Incident response is an organized approach to addressing and managing the aftermath of a computer security incident or compromise with the goal of preventing a breach or thwarting a cyberattack. Incidents that are not identified and managed at the time of intrusion typically escalate to a more impactful event such as a data breach or system failure. The intended outcome of a computer security incident response plan is to limit damage and reduce recovery time and costs. Responding to compromises quickly can mitigate exploited vulnerabilities, restore services and processes and minimize impact and losses.

Incident response planning allows an organization to establish a series of best practices to stop an intrusion before it causes damage. Typical incident response plans contain a set of written instructions that outline the organization's response to a cyberattack. Without a documented plan in place, an organization may not successfully detect an intrusion or compromise and stakeholders may not understand their roles, processes and procedures during an escalation, slowing the organizations response and resolution. There are four key components of a computer security incident response plan:

Preparation: Preparing stakeholders on the procedures for handling computer security incidents or compromises.

Detection & Analysis: Identifying and investigating suspicious activity to confirm a security incident, prioritizing the response based on impact and coordinating notification of the incident.

Containment, Eradication & Recovery: Isolating affected systems to prevent escalation and limit impact, pinpointing the genesis of the incident, removing malware, affected systems and bad actors from the environment and restoring systems and data when a threat no longer remains.

Post Incident Activity: Post mortem analysis of the incident, its root cause and the organization's response with the intent of improving the incident response plan and future response efforts.[12]

Sanity Check

TechRepublic's Jason Hiner points out how traditional security models are breaking down and we need to shift our focus to protecting our data. Hiner spoke with IT security experts Kris Lamb, Director of X-Force Internet Security Systems for IBM, and John Pironti, Chief Information Risk Strategist for Getronics about the future of cybersecurity:

So if traditional perimeter security will no longer work, where does that leave us? Lamb and Pironti both pointed in the same direction —> data security.

Pironti said, "Let's define the new perimeter by thinking about how our data can be impacted, not how our technology can be impacted. That is the biggest challenge. Security professionals still run to the box. It's still too easy to just buy the box...The perimeter is wherever the data is. The perimeter follows the data."

Lamb said, "Data security is a solution that people need to be thinking about... Data is becoming the delivery mechanism of a lot of the real nasty threats that are out there. You're seeing trusted file formats [DOC, PPT, PDF] used as ways to embed malware and exploits."

So what's the difference between perimeter security and data security? Pironti contrasted it as the difference between building a big castle with a moat around it versus building really strong body armor for the knights you send out onto the open battlefield.

But the approach to data security is also about much more than just using different tools to protect smaller targets. There's also a philosophical and cultural shift as well. "The most important thing that we think people should do is to do a threat vulnerability analysis," Pironti said. "Let's look at your data and your business processes, and not just your technology. Let's look at your information infrastructure, which is all of the people, processes, and technology associated with information and data. Now, start looking at that and saying, 'What are the possibilities in how that data could be compromised? What is the likelihood of that happening and what is the potential business impact?' So you describe all the scenarios and lay out all the possibilities... Once I understand what can happen to me... then I can start talking about what is my vulnerability management plan... What I am going to put in place that will still enable the business to function but protect it from these potential high-threat, high-likelihood situations? That is a process-oriented, business-oriented approach to information security that hasn't existed in many organizations. It still doesn't today because it's too easy to just buy the box"[13]

Everything is rapidly changing in the digital age, like mobile applications, social media, and all related applications, virtual goods, and of course, the wild web-based information management and marketing systems existing today. I want users to master the situation and take action for yourself, family, and your business today.

11

STEPS TO SECURING YOUR LIFE AND BUSINESS

"The future is not preordained by machines. It's created by humans."
—Erik Brynjolfsson

Clean Up Your Technology

Cybersecurity consists of preventing, detecting, and responding to cyber confrontations (incidents) that can have widespread effects on the individual, organizations, the community and nationwide level.

With an Unhackable mindset, you can conquer attacks on content data and minimize threats against your digital identity. Are you ready to take the steps necessary to stop the unwanted invasion into your private life?

Your content data has become a high-value target. The following steps will help you set up and restore your online defenses. With practice and discipline, you can maintain security with every connection. True security starts with the user.

Give your private and business data relevant categories so you can use and protect your data more efficiently. How you classify your data is up to your personal and work environments; this can differ from business to business and user to user. We must classify our data so we can create a proper setup of our personal and work environments. This proper set up will help reduce certain security and privacy problems like duplicate pieces of information stored inappropriately, misplaced data files, and reduce clutter, which will increase device productivity.

Once the correct thought process is realized, you can create a safe and secure operation. You will see long-term privacy and security layers fall into place as you create your true, real-time, end-to-end psychological defenses.

Having an Unhackable mindset is about the correct setup, configuration, and installation of privacy and security controls that you put in place. Remember, it took many years to accumulate your online data, so it will take a bit of time to clean up. Take your time with this chapter and recognize that you will not complete all of these steps at once. Each step you take will bring you closer to an Unhackable mindset and true data security.

Phase 1—Prevention

Hackers first ask themselves this question: How will I attack this system?

The first phase is understanding that your current setup must happen before you dive into the world of data protection and management. You must first know what data you are dealing with to know and better understand where your data is going, before you input and execute any remote commands. This involves taking a detailed look at the inner workings of your personal and business data and how you truly use it, or how it could be used against you.

When starting any project, you must start off properly. Benjamin Franklin is credited as having said, "By failing to prepare, you are preparing to fail." We always need to take a good look at what we are going to be dealing with before we do anything with it. We need to assess what information we have that is private and what is public, so we can get a comprehensive view of all our data. This new understanding of our set up will help us in navigating all our information and setup to increase our data integrity check and balance systems.

Focus on knowing your data and begin with a fresh computing environment by purging all unnecessary apps, data, etc. from your computing environment. Begin to assess and better understand your IT needs so you can ensure your personal and business technologies run as effectively and securely as they possibly can.

The following common issues and characteristics should not exist in your personal or business environments:

Reconciliation Issues: New technology comes out constantly. We must constantly plan and assess what our technology needs are and if they are being met. When implementing new technology within our existing workspace, sometimes old doesn't play well with the new. We must thoroughly vet any new technology we want to add within our workspace and do the research to see if there could be any issues with legacy (older) devices that are still crucial to our personal and business day-to-day operations. This means putting the work in ahead of time so that we don't have to spend time worrying about any future problems that could have been solved with a little research and a different purchase.

Another key factor to proper integration of new devices in the personal and work environment besides vetted research is proper set up of these new devices. This is called "zoning" a computer. This means removing any and all third-party programs or "bloatware" that comes pre-installed on the device. After the device has been purged of all irrelevant programs and data, we need to then configure the device into our existing personal and/or work environment, such as proper IP configuration, mapped drives, peripheral device set-up and confirmed functionality (i.e. legacy printers working on new devices).

Alternative and Substitute Readjustment Challenges: When we are working in our day-to-day lives, we never think of what the worst-case scenario could be, that the next day we might come back to work with a technological disaster. Power outages/surges, server crashes, theft, ransomware, or any other unforeseeable event could result in a partial or total data loss. This is the moment we think to all the times we should have implemented and maintained a data backup system. These systems are essential in any personal and work environment.

Implement one right away and establish a routine of a backup policy so that you can prevent any data loss or business downtime. A good way to establish the routine of a proper backup system is using proven backup programs that help automate the process, like Back4All or Data Deposit Box. Both programs backup selected information, whether it be offline or online respectively; however, it is always preferred to back your data up offline so it can remain secure.

Foundational Infrastructure Cracks: We need to ensure that when we are solving an IT issue, we are removing the cause, not a symptom of the cause. The extra thoroughness we need to implement in finding the root cause of our problems will save time that would be wasted in the future retroactively fixing symptoms. Finding a trend or pattern is crucial to establishing the root of a problem. Therefore, it's vital for users to keep track and monitor any problems they have, so that their IT professional can fully understand the cause of the underlying issues.

As a user, you need to become a detective. You need to investigate and follow the trends that create these problems so that you can prevent them from happening in the future. Being a detective means being extra vigilant while using your computers so you can quickly identify non-routine behavior. This self-empowerment will help you be more confident when addressing problems with your devices. The majority of IT problems are user error and can be resolved with a little investigative work by the user.

Contingency Safeguard Measures: Now more than ever, hackers roam the Internet looking for their next potential victim. With the creation of smartphones and other modern-day devices, taking advantage of someone with relatively public information is easier than ever. Scams are becoming more and more sophisticated due to our private information becoming more and more transparent. Therefore, we need to establish a layer of defense around us and our private and public data. In personal and business settings, this can involve many different things:

1. Physical IT security equipment

2. Anti-malware and antivirus software

3. Establishing security protocols

5. Intrusion prevention software

6. Security patch management (Ensuring OS security patches on devices are up to date)

7. Secure backups

8. Multi-factor authentication (Can be enabled in many programs and services used in the personal and business settings)

9. Security Awareness Training (Scam Awareness Training, Active Education in Security Risk Assessment)

All of these strategies and tactics should be implemented while creating a secure personal and work environment and limiting security risks.

Lack of Strategic Data Processing and Storage Planning: As personal lives get busy and businesses develop and grow, people tend to let their IT support fall by the wayside. You must constantly keep your IT departments in the loop and have them keep you in the loop when it comes to planning the expansion of your business and its IT needs. By keeping IT in the loop, you can proactively assess what future problems or new devices you will need. This give and take will let everything run more smoothly as you are aware of your growing needs and concerns, and your IT department is not thrust into a new and unprepared environment.

This could involve your IT department with infrastructure changes you will need to make to expand or what security advancements will need to be implanted to keep your personal and business secure. Being in the vast ocean of the IT world can be daunting, but proper planning can help reduce complications down the road.

Expecting the Unexpected: Most people tend to wait to report something to their IT department until they physically can't operate their device anymore. This is a huge problem in preventing major downtime in personal and business settings. You need to be actively aware of how your computer is supposed to operate so that when a miniature problem or pop-up occurs, you can figure out immediately how to resolve this situation. By keeping a step ahead of the problems, you can solve them and prevent any future problems from occurring as it becomes much easier than if you were constantly playing catch up with your mistakes.

Discouraged Users: When any user or employee begins to realize their computer is slow, frustration is not far off. We assume that everything we do is immediate and must be lightning fast. This disconnect from how technology works to how we want it to work is created from a lack of understanding. We need to educate ourselves on how we can streamline and expedite our day-to-day operations. Once we have done this, we can look at the technology and see if it can be updated or upgraded to better suit our needs.

Phase 2—Patterns, Protocols, and Policies

"Nearly all men can withstand adversity, but if you want to test a man's character, give him power."

—Abraham Lincoln

"With great power comes great responsibility."
—Stan Lee, *Spiderman*

Hackers next ask themselves this question: How will I interfere with its operation?

The second phase is the configuration period where you create your data classification policies. There is a regular *rhythm* of usage employees have with respect to their most used apps and high value activities. These activities may be common among teams and require app-specific training with respect to creating a safeguard. In those environments that have a rhythm, there is a place for becoming mindful of how they are affecting one's attention.

The following steps will help you be able to detect where your data is being transmitted and used while it's happening and processing in real-time, during all confrontations and incidents. Once you have established what data you have and whether it's public or private, you must create a data classification policy protocol that you can follow in any given circumstance when it comes to your data to protect it. How you choose to interrupt different levels of private information is crucial. When you decide ahead of time what to do and how to handle your information, it creates a better flow within your personal or work environments. You are spending time before you do anything to save time later. As the adage goes, "Measure twice, cut once." By taking the extra step of measuring twice, you eliminate future problems that could occur down the road from mismanaged data.

While working during personal or business hours sometimes we can't afford the luxury of taking missteps with public and private data. Much like cutting wood, even if we mishandle confidential information once, the steps we take can sometimes be irreversible. Therefore, it is important ahead of time to create a data classification policy so that, while we are handling public or private data, we automatically know what to do with our data and how to manage it properly. These policies will limit frustration and mitigate potential security risks when it comes to our data.

Combined Characteristics for Privacy and Security

Avoid Data Sync Programs: Don't let your data evaporate into the cloud. If you can easily access it, so can anyone else. There is no such thing as a truly secure system in today's day and age. Having all your data stored in one location means your digital identity can be accessed by anyone; it's not good to throw all your eggs into one basket. If the information you're putting out there is important to you, ask yourself if you would want someone using it for their own personal gain. What happens to that one location which has your entire life stored in one easy and convenient place or smart device if there is a data breach?

Establishing an organized influence over your data is key. With constantly changing user interfaces, it becomes increasingly challenging to go about installing and using programs securely and as intended. Data syncing is bad because it creates a new complexity as it throws data everywhere on the cloud and on different devices. This complexity will become your adversary, and it will develop a masked domino effect that puts you and your data at risk. We must begin to limit the amount of data we actively choose to expose and thus mitigate our risks.

Protect Your Big Data: Personal information is being bought and sold quietly. Control your environment because it's where you are receiving the information. Companies and third parties survive off your data so much that they design Customer Preferences, like built-in Convenience, Effort, User Interfaces, Communication & Information, Stability vs Variety, Risk, Values, Sensory, Time, Customer Service, and Customer Experience similar to a maze to try and slow you down from changing their predefined, default variable settings. It is like a game of Mouse Trap with the company's products and services as they don't want you to escape the predefined triggers.

Being vigilant of what default settings are turned on and how they will impact your privacy and exposure is the key to securing your private information. If we are the mouse in this game, we need to think about how we can escape and break these conventions. We can't make a plan of attack if we don't know who or what we need to defend ourselves from. New triggers are always ever-evolving and changing.

Carefully Install Programs and Remove Bloatware Software: Before you click yes or continue, make sure programs aren't trying to sneak unwanted software onto your devices. These installed programs have become clever at confusing users so that they are not fully aware of what they are allowing. It creates a chain reaction by putting pesky, undesired software on your device. These programs also have complex user interfaces and buried settings deep under many layers, just so that it becomes more challenging to uninstall. Don't let minor irritations morph into full-blown infections that could have been avoided with patience and thoroughness.

Backup and Archive Policies: Having easy access to your data online makes it susceptible to hackers. It is crucial to archive all your data offline locally on an encrypted external hard drive to keep it secure. If your data is locked away in an offline location, then the only person with access to it is you. It can be a hassle to constantly move information onto hard drives, but if it is away from the Internet, it is away from harmful threats. You'll need a backup of the main external hard drive while keeping it in an undisclosed location as you rotate them on a set schedule. Test your backups often. These preventative policies we put into our data backups will save us potential trouble or loss. As Benjamin Franklin once said, "An ounce of prevention is worth a pound of cure."

Limit the Use of Free Platform Environments like Google

and YouTube: When using free platforms, there is always a risk of your personal data being revealed through legal and illegal tactics. They can sell your data. Nothing in life is free. These platforms have huge traffic numbers on their sites, which means a large base to target ads to. When agreeing to use these sites, you automatically sign away most of your rights. Be careful with what information these sites have access to and limit what information they store so you can avoid systematic analysis of your data or statistics. A good example would be Google Ads, which "is an online advertising platform developed by Google, where advertisers pay to display brief advertisements, service offerings, product listings, video content, and generate mobile application installs within the Google ad network to web users."[1]

Short-Term Gains or Rewards to Stay Away From: Ease of access tools help us save time and everybody does it, so they must be okay. Actually, this technology you think helps to simplify your life makes it more complicated, because there is a whole new group of trackers in your life keeping tabs on your data activity. Remember, your short-term fixes yield long-term problems.

Watch out for these short-term gain online activities:

- Detached Social Interaction (We connect, but not a deep meaningful human interaction.)
- Fun/Enjoyment (The rush of endorphins from games and apps, but feeling old if you aren't on the latest app.)
- Purpose/Time Waster (Can give us something to do and help avoid boredom, but we need proper priorities.)

Think Twice before Using New Technology: There is no such thing as a "free lunch." Always be cautious of what third parties may want from you in return for their platforms.

Whether it be to sell you ads, services, or even your own information, never forget how valuable you, the user, are. Innovation is infectious.

Distinct judgment is required in everything you do with technology before any use or experience. Like merging onto the highway, you need to yield before you access any main route.

Technology is only as good as the user who uses it. Slow down when using any technology; it's more than a caution sign and more like a yielding process that becomes your defense in depth.

Disable application access to any camera, microphone, location, and any privacy features in your configuration settings for all Siri, Cortana, Alexa, or other voice-enabled options. Make sure to also change and set any "always open" permissions and set to either "never," "ask next time," or "while using the app." As a rule of thumb, you should never have anything set on "always" or have unrestricted, open access for anything if you're looking for security and privacy. Having a listening voice there to simplify your life actually makes it more complicated because there is a whole new tracker in your life keeping tabs on your private data, usage, and activity.

Familiarize yourself with the configuration settings for hardware, software, apps, and services.

Read Over Agreements: Agreements are designed to be complex and wordy to get you to give up and just say yes. Taking the time to understand what you are agreeing to means you are securing long-term peace of mind and avoiding short-term headaches. If you don't feel comfortable, then don't agree to anything. At a minimum, at least scan it for anything that raises a red flag.

Principle asset-based decisions and choices need to be made regarding what kind of countermeasures to enforce with the information you provide to cookies, web beacons, and other tracking technologies, third-party cookies, third-party web beacons, and other third-party tracking technologies, mobile device and mobile browser information, location data, usage logs, calls and text messages, public content, and social networks.

Focus on Verification and Validation: Don't rush to fix a problem without understanding its existence to begin with. You are your best security measure, so instead of asking why something failed, exercise restraint to ponder where you may have made an error in judgment. Large-scale security issues often stem from the inability of users to tend to small nagging issues, which in turn evolve into debilitating obstacles that require professional hands to mend.

Start by asking questions and giving yourself time to address each issue without it all compounding to create a bigger mess. Doing little things well is a step toward doing big things better. We can't build and expand upon a poor foundation. Our workflow depends on the constant building of what already exists. If we have an unstable foundation, we will inevitably collapse. When solving problems, we must always consider the source; we must cure the disease, not just treat its symptoms.

Don't open every attachment. There can be danger lurking in what appears to be harmless attachments. If you aren't familiar with the source an attachment came from, don't trust it. Only open attachments from senders you can verify. Otherwise, you are painting yourself as a target to attack. If hackers think they have a foot in the door, you can bet they will try and push themselves all the way through.

The most common thing I see is when users never close their task at hand or restart or refresh any device. You need to make sure to always manually or automatically close tabs, apps, and programs after any usage. At a minimum, set the option to close after one day, one week, or one month. For example, I see devices running without a normal reboot or a shutdown refresh for months and months in a row. You should never keep anything open forever, as this will create an immediate and present danger and create many more security and privacy risks.

Whenever you're executing any task, you need to know what it's doing or what it's changing behind the scenes before you run anything in an automatic style cleaning, fixing or scrubbing manner. You should always find the root cause before you move on. You don't want the underlying issues to mount on top of one another creating bigger issues and outages or moving over to your new system setup during system migrations.

Stop Smartphone and Device Tracking: Configure settings on your phone and apps so that companies can't track where you are going. If they know where you're going and have been, they may even try to suggest your next location. This is due to third parties letting advertisers have access to your data in order to sell you their products. Turn off or pause device usage and activity in your data settings to protect yourself from having decisions made for you. Make sure your smartphone is not programming you, but that you're programming it. Be aware that your smartphone knows more about you than you do. It's a warehouse of private information, so use it with caution and secure your configuration options with care.

Avoid location services where companies want to track your location in order to sell you ads and services based on where you are and have been. Ever notice how some ads seem just a little too convenient? Now you know the reason. Disable this to remain free from outside influences and in control of your decision making. Always keep your locations private and off.

Are there any GPS apps that don't use data? We have become slaves to our cellular providers' data plans. Many of the leading GPSes work only when they are connected to a network. The solution to this would be a GPS with offline map support.

Be vigilant and aware that these smart devices are like scanners. They are scanning all your information on your phone. Even if you haven't saved any contacts, it's scanning and searching personal emails for contacts and associating them together habitually.

Improve Password Practices: Devising only one challenging password with various letters, numbers, and characters at its core is basic protection. As hackers become more sophisticated, so must our password practices. Choosing different, vague, and complex passwords for each separate account and changing it regularly is a necessity. Some techniques to try are using the beginning letters of a sentence you will remember or using an anagram that you will remember.

There needs to be at least another unique layer of protection to frustrate hackers. This is known as multi-factor authentication, which involves a verification of your identity when you input your initial password. Secondary verification includes specific locations and actions, biometrics, or even a physical token. These complicate the hacker's efforts, so they will move on to the next target.

Make sure to change all your app, services and device default passwords. Set up passwords and encryption layers of security onto your hardware and software, like your BIOS. Enable enhanced password policies, set user screen timeouts, and limit user access. There are now dark web research services that can scan the dark web for stolen credentials that have been posted for sale.

If you have a very simple and common password that's seven characters long ("1234567"), a professional could crack it in a fraction of a millisecond. Add just one more character ("12345678"), and that time rises to five hours. Nine-character passwords take five days to break, 10-character words take four months, and 11-character passwords take ten years. Build up to twelve characters, and you're looking at 200 years' worth of security. Another factor to consider is time. Passwords grow weaker as technology changes year to year, and hackers become progressively skillful.

Avoid Public Wi-Fi: Your data is actively exposed when accessing public Wi-Fi. To safeguard your data, it is best to stick with private Wi-Fi connections. However, if that is not possible, a paid VPN (Virtual Private Connection) can cloak and encrypt your data. Make sure it is encrypted on both ends, so hackers can't find a way to slither into your connection. In doing so, your privacy remains unscathed by those who seek to exploit it. Spoof your hardware MAC address for extra security. Disable and prevent Wi-Fi from automatically connecting to open networks.

Security Assessment & Education and Training Awareness: It's important to dedicate time to establish a plan of attack and have a baseline budget of your resources that are going to be spent. A good way to start is to ask yourself, when was your last assessment?

Train your users by educating them about data security protocols frequently through different policies, such as a Written Information Security Plan, Termination Policy, Security Incident Procedures, Sanction Policy, Network Security, Access Control, Computer Use, Disposal Procedure, Bring Your Own Device (BYOD) Policy, and Facility Security Plan. Collectively, they work toward the three fundamental principles of security, which are confidentiality, integrity, and availability. First, by ensuring the security and confidentiality of data accumulated by and in the custody of the Organization. Second, by protecting against potential threats or dangers to the security or integrity of such data. Finally, the availability ensures dependability and timely access to data and resources to authorized individuals.

Use a Better Email Address: Use at least three different email addresses. The first one to receive emails from important sites and apps, the second one to receive emails that aren't so important, and lastly, the third one should be a security reset and recovery option that is from a different email provider to avoid having everything in one place hacked and losing access to your most important activity.

Using multiple email addresses can help filter out important information and documents from the mountains of junk sent from third parties and other malicious sites. There are free email domains, and there are paid ones. You might wonder how the free ones can be free and why the paid ones cost any money at all. We pay for what we get. By putting in the time and money to commit to a paid and secure domain, we have already increased our level of security. It allows for users to interact with new sites while keeping a barrier of protection in place from unwanted hazards.

Having a secondary email acts as a practice dummy, allowing you to decide if a site is worthy to have access to your primary personal account. This extra barrier provides space to breathe so users aren't overwhelmed by the growing amount of spam many inboxes have fallen victim to. Secure your email because most attacks start there.

Use a Flip Phone, Light Phone, or a Custom Rom on Weekends: The benefits of using a "dumber" phone allow you to disconnect from the constant hailstorm of notifications while retaining the basic important functions we originally needed our phones for. It doesn't have to be permanent. You might only use it occasionally. However, it creates a space for people to catch their breath from the suffocating world our ever-evolving tech puts us in.

Update Your Technology: Making sure software and hardware are up to date is crucial to combat new threats. However, be aware that with each update companies will try and sneak in unnecessary "free stuff" for their nefarious ends and not for added protection. Only take what is useful and practical to maintain your security. Turn off all feedback, diagnostics, and any other settings to limit access to what data is being sent out routinely. Always check for critical updates on all devices.

When upgrading your software and hardware, you need to step back and look at the bigger picture. You need to see what is different from before and why. You need to become a detective to figure out the motive behind these changes. Are they beneficial? Do you need them? Can you continue to function without them?

One of the most commonly missed steps when using the cloud or local technology is the belief that you don't have to do any routine maintenance regarding your technology because it does it all for you. That's incorrect. Just like your lawn, your technology needs constant care to stay healthy and look its best.

Delete Activity Controls: When logged into an online connected account, the connection acts as a conduit for transmitting your online browsing and activity. Don't let that information be gathered by managing your data and personalization settings, clearing it, then pausing the settings. By doing this, you can feel freer without having to worry about how your searches are being accumulated and monitored. Make sure to always disable and shut off all privacy settings that share any personal data or activity.

Don't Let Any Products or Services Give You a False Sense of Security: Apple has a wealth of knowledge and years of practice when it comes to providing airtight security on their devices and software. However, they are not invulnerable. Just like every other major company, they have breaches and will never be able to provide bulletproof security. Always question what security promises are fact-based, and which are optimistic expectations from large tech companies.

The answer is not found with any Open-Source products and services like Microsoft, nor any Closed-Source products and services like Apple, but in you. You are the main defense for your online and offline security. It's up to you to put in the time and energy into finding out what is and isn't secure, and what practices and policies we need to implement to keep ourselves and our data safe.

Don't Fall for Spam Phone Calls and Text Message Scams: Unfortunately, it is difficult to avoid the barrage of mysterious phone calls and text messages our mobile devices receive. The easiest solution is to ignore and not engage. If you engage with any foreign numbers, you are only incentivizing unwanted threats to continue their assault on your device. Many of these numbers are typically ads or act as a backdoor entrance to bypass your device's security. As a rule of thumb, don't pick up unknown numbers; always let them forward to your voicemail.

If you are using any iCloud solution, don't allow any critical or private information to be left on voicemail. It is being stored online and backed-up in a remote location accessible to be restored under certain situations. Hackers can listen and use different pieces of information sitting in your voicemail box to put two and two together for a larger task at hand.

Be Careful When Using Apps and Trusted Devices: With the ever-expanding number of apps available, so grows the number of holes in their security features. Many of these apps share information with one another, so that if one gets hacked, the rest are now vulnerable. Due to this, caution is always recommended when installing new applications and understanding what permissions they have access to. Some apps are just fronts designed to steal your data and wield it for their own financial gain. Be selective in what you download and allow on your devices and don't connect one app to another.

Make sure to always clear old devices and try not to trust any new devices. It would be better to authenticate on-demand every time than be careless with your security and privacy.

Use Facebook and Other Popular Social Media Sites Carefully: The Internet is permanent. You may think you have deleted something, but someone can quickly snatch and save that information as soon as it is available online. It will also be archived by the Wayback Machine, which is a digital archive of the World Wide Web, founded by the Internet Archive, a nonprofit organization based in San Francisco.

When you post, ask yourself if you are comfortable with that information living online forever. It may seem like a dramatic approach to posting, but it is the harsh reality we find ourselves in. There are always those lurking in the shadows waiting to seize your information for diabolical purposes.

Be careful with social apps like Facebook Messenger and other messenger programs and apps. Set apps to use minimal permissions, so they can't access various aspects of your phone. Allowing access to your camera, microphone, location, and similar features paints a detailed picture of your actual identity without you knowing it. Don't let companies and third parties get a full portrait of who you are, because then they are able to tailor specific ads and services to target you. You should review your configuration settings end-to-end. You will instinctively know what to choose as you comb through it.

Focus on protecting your public information, keeping your private identity secure, and all its data content associated with every use. Limit access and permission for ad settings to be more random. Also make your profile private to your friend groups instead of being wide open to the public. Make sure to restrict anyone from posting on your page without your permission and approval every time, protecting your identity from anything that can go viral.

Phase 3—Prioritize and Organize

Hackers ask themselves this third question: How will I destroy its integrity?

The third phase is focused on prioritizing and organizing your data and knowing where your data is stored in transit and at rest. Now that you know what data types you have and what policies you want to enforce to properly manage your data, you can begin to prioritize and organize data and start to physically organize your data within your unique criteria.

You can start tagging your data based on how private or sensitive it is and create a secure foundation for your information. By tagging your information or marking it within a system that you can understand within your personal life and business, you can begin to create an environment, whether it be physical or digital, that naturally defends your information from mismanagement. While your data remains organized and easily accessible to you, at the same time, you create layers of defense.

Proper data management policies, like consistent and secure tasks and organized personal and workspace environments, reduce exposure to your private, sensitive data.

Proper data classification and organization is a cornerstone to having a secure, productive, and distinguished personal and work environment. There are several routines that should be established to create this type of environment. One routine is reducing clutter between your personal and work environments by purging old and irrelevant files. A helpful tool for this is Cleaner, which can help in reducing clutter and removing junk files within a user-friendly program.

Using compliant and secure archive and backup systems will help create a secure and scheduled redundant personal and work environment in case of data catastrophe. Taking the time to educate yourself on how and what data belongs to your personal space versus your workplace needs will help manage your private and public data.

The Beginning of a Transformation

By beginning to implement these security and privacy steps, you are well on your way to establishing the Unhackable mindset as your only secure environment that will clean up and course-correct your digital life. As you continue to secure your technology, you also want to develop a healthy mindset around technology.

The outcome will cause a ripple effect in both business and personal environments immediately. A real understanding of cybersecurity will empower you to share that education with friends, family, business associates, and personal colleagues. The possibilities are endless.

12

THE UNHACKABLE MINDSET

"It is not the strongest of the species that survives, nor the most intelligent that survives. It is the one that is most adaptable to change."
—Charles Darwin

More important than the external work you do to secure your digital devices is the long-term, internal work you need to do to shift your mindset. Begin to transform your mindset by reflecting on and implementing the following skills and challenges.

It's important that people know how much of themselves they are giving away by not securing their internal resources. Most people don't understand that while it seems like little bits of data, they would be shocked at the amount of privacy and security they are losing without knowing it.

Whether we like it or not technology is here to stay; the digital age in upon us. Just as we went to school to read and write, we must also educate ourselves on the principles of intelligent technology. Illiteracy is a personal choice; algorithm literacy is also a choice. If you want to become part of the digital age, then take the necessary precautions. Learn to leverage technology so it works to your advantage.

Companies have been creating innovative technology, while pushing the thresholds and limits of users' understanding and knowledge. Now most of the same companies are trying to sell back a solution to secure our information. However, these companies can never truly provide security and privacy. Each user has to take proactive steps and empower themselves.

No one is ever truly safe. It does not matter what software you have loaded on your computer, or what product or services and programs you purchase, everyone gets hacked at one point. The only difference is how you set up and prepare to handle the situation before any signal is sent out or received. The technology industry is in need of a genuine new approach or model to IT.

The only Unhackable software is the human mind. By educating yourself on the basics of internet usage, algorithms, and privacy, most breaches can be stopped before they happen. Let's use the internet to our advantage, rather than the internet using us.

"As human beings, our job in life is to help people realize how rare and valuable each one of us really is, that each of us has something that no one else has—or ever will have—something inside that is unique to all time."
—Fred Rogers

Privacy and security are not dead, but are in a struggle with third parties who are not only vying for control of our privacy, but the narrative as well. We all need to grab our own narrative. We need to get our foothold in before bad actors distort our concept of what privacy is.

"I feel so strongly that deep and simple is far more essential than

shallow and complex. In the end, life isn't about material things. It's about the relationships you have with one another and yourself."
—Fred Rogers

Deep and simple applies to relationships and conversations. If in any corner of our lives, including our sacred spaces, we deny the opportunity for a deep and simple experience, we are truly neglecting something great.

Your journey to achieving an Unhackable mindset is all about setting the groundwork through developing new skills. A skill is "an ability and capacity acquired through deliberate, systematic, and sustained effort to smoothly and adaptively carry out complex activities or job functions involving ideas (cognitive skills), things (technical skills), and/or people (interpersonal skills)." As your frame of mind transforms, it resets your new baseline norms and values by utilizing the greatest technology ever created—the human mind.

The Ladder of Internet Access

Connecting + Sharing + Growing = Transformation of your true genuine self.

Part 1: Living in an Unsecured World

Skill 1: Rejection

In the initial phase of rejection, users will learn to dismiss their current knowledge of Internet usage, thereby abandoning harmful situations and creating space to learn new ways of handing their digital data. *Unhackable* teaches how to expand on the good aspects of Internet usage, such as the beneficial communion of healthy connections with colleagues and maintaining rewarding friendships, or developing long-lasting, solid relationships. The spirit of self-control develops from an inner strength used to conquer the problems acquired through Internet usage. By developing the leader you were designed to become, those addictive properties of the past will be eliminated.

Skill 2: Disconnection

In order for this next step to be successful, you must detach from the dream world of the Internet. We live trapped within a system where every action creates a reaction to keep us incarcerated by data-driven technology designed to steal our personal information. Once withdrawn from the snare, we can become leaders not followers. The scene clears a path led by peace; enabling you to walk a road free from digital systems. Regain control of your data privacy freed from the erroneous intelligence of the Internet.

Skill 3: Discretion

Discretion brings forth a perfect balance between our private and public information. The choice of freedom to decide what should be done in any situation online dealing with technology. You can behave in such a fashion to avoid revealing private information. Stop the addictive nature and separate yourself from the draw of social media. We can be humbled by the realization of our insufficiencies.

Part 2: The Foundational Integrity of Freely Shared Information (The Fundamental Virtues)

Skill 4: Compliance

Safeguarding your personal data in this digital society requires compliance with certain established guidelines. This mindset will allow a leader to excel in all areas of importance. When a person concedes to following a set of rules designed to protect personal priority information, the end result establishes dominance over the hackers trying to steal your most private data.

Skill 5: Self-Reproach

With this skill, we realize shielding private data lies solely at our doorstep. We are accountable for this transfer of power over personal data. But, with daily amendments, we can achieve dominance over the vices of data control.

Skill 6: Eradication

The next phase with your Unhackable mindset is the process to eradicate the control over our connections. We must completely remove the destructive influences in our lives. Once these governing powers are eliminated, the gateway of the mind will be revealed. It grants total power over technology data breaches. You will learn how to be in the grid, not of the grid, or on the web, but not of the web.

Skill 7: Data Cleansing

Of all the skills in this process, data cleansing is the most satisfying. When you learn what information is good, the harmful nonsense and the most damaging data must be removed. Developing a quality data plan with clear expectations for your data is a must. Standardize your data review at the point of entry so that you develop the mindfulness skills to validate, identify, and purge data before it becomes damaging. Following this standard operating procedure will ensure that before cleaning data even happens, you are checking important data at the point of entry. As you routinely review your data, your Unhackable mindset will begin to more confidently govern your online experience over the Internet. You will become more aware of all the unsafe hotspots related to your connected online experience.

Skill 8: Uninhibited

By now, you have learned the importance of proper Internet control. The cravings for relentless Internet connection is past and your code of ethics has developed a moral standard. Henceforth, the connections with data remain naturally organic.

Part 3: The Powers Hidden Under the Surface

Skill 9: Malice

Next is the removal of all negative data influence. We can continue to embrace our freedom without the impediment of clutter on the main purpose of connection. Continuous clicking generates data to feed the algorithms, which becomes your Internet identity.

Skill 10: Character Assassination

Unhackable helps to maintain free speech that is not used to propel corporate agendas. Damage to a person's or company's reputation can be limited. When these statements are created for malicious intent, it stalls progress and wastes valuable resources. Problems are not resolved in the solution, but rather with an understanding of how the issue was created to start. Temporary fixes do not eliminate the underlying cause; they only serve to extend the problem. Instead, build a foundation of integrity.

Skill 11: Communication

Technology moves faster than most people realize. Data is being replaced with more data and so forth, until the situation reaches a tipping point. We cover the problem with a quick fix that only disguises the underlying problem. Like a contractor just painting over a mold spot on the ceiling, it would be smarter and safer to repair the problem. Once the initial damage is done, it's difficult to reverse the effects. The Internet platform is constantly changing to keep people insecure. When you realize the problems with these new innovations, it's easy to step away to find security and peace of mind.

Skill 12: Fabrication

The Internet is filled with fabricated data collections to keep you captivated. It is manipulation to maintain a steady click rate from one area of the Internet to the next. All the while, choices are being collected and gathered to create your data identity. It's imperative that you see the larger picture instead of one mouse click at a time.

Skill 13: Boredom

Habits often appear from deeper issues. Stress and boredom are just the residual effects. These problems are tough to acknowledge, but change requires commitment. Be honest with yourself.

Individual beliefs bring these practices to the surface without the person's conscious understanding. Fear and anxiety often produce debilitating effects. The paralysis creates an ineffectual way of living in relation to the Internet. We must liberate ourselves from the incapacitating effects of the web. Once the chains are broken, freedom will develop an unforgettable, liberated state of mind.

Skill 14: Lack of Awareness

When we become desensitized, our response to change becomes non-existent. People's actions develop into something robotic. Our thought patterns and actions are based on instincts alone. Society moves forward as a component linked to the system. The options to escape captivity are limited because technology is moving at the speed of light. Technological advancements were developed as a tool to make life easier. But it has become a distraction we use to escape reality. We have lost the concept of balance. Society is powerless without technology. Negligence becomes a habit. Thoughtlessness creates a trap, and ignorance becomes a gateway for problems.

Skill 15: Sleep

We must learn how to gain control of our mental states. When our data is strategically stored and used against our wishes, then there isn't a natural flow of life but rather a preemptive strike against our personal choices and freedoms. Are you in control or is it the data derived from your life in charge?

The capturing process of collecting our data happens when we sleep. When we are unaware of the happenings of Internet usage, ineffective ad-hoc connections create an unstable foundation, which data miners can use to steal your data.

A sleep-deprived brain is like hacking your way through a dense jungle with an axe. It's overgrown, slow going, exhausting. The paths overlap, and light can't get through. A well-rested mind is like wandering along a beach with gentle waves caressing the shoreline; the paths are clear and connect to one another at distinct spots, everything is in place, and you communicate accurately. It's invigorating.

We should not allow our privacy to rest in someone else's hands. Do not trust the product or the service of the Internet but rather have faith in your instincts to achieve pre-planned goals.

Do you sit behind the wheel of decision making, or is the information cultivated from you?

Skill 16: Alertness

By this time in the process, you must be aware and ready to escape the reality of data mining. You should welcome the freedom and be ready to act upon the commands set forth in the previous steps. It's imperative we are part of the system, but not one with the system. We must remain vigilant and actively squash any further invasions of our privacy. Methods of Internet data mining come in many forms and from all directions. A program's intention must be judged and allocated for proper usage.

Skill 17: Fear

One secret defense in data mining is to instill fear by proclaiming an event that may never happen, keeping people apprehensive. It may be something simple, but it is made to appear dangerous. Urgency causes someone to react unconsciously. True security is obtainable once you understand how the game is played. For example, antivirus software is completely useless as cybercriminals are aware of loopholes. Attacks cater to those weak links, the human factor. Most malware and ransomware viruses are spread by users who fall for phishing techniques. The action is a proven source of fear porn. Panic is not so much a matter of a click but an incarceration of the mind. An attack will focus on what is not known, so you remain distracted and the assault will work on fear. Fear is the loss of assurance. When the comfort level is kept to a minimum, you must continue to move, adapting to the changes. Therefore, the danger of losing your data is maintained as a threat.

Skill 18: Extreme

Technology can eliminate productivity. Excessive use of anything can end up controlling every area of your life. To the point, connections maintain power and you become a product of your actions. The concept is to steal your real identity. Technology was created to free us, not become our prison. We must take action to break away from the imprisoning effects of Internet usage. Society must find a common balance between old school methods and new age technologies before our foundation erodes to the point of no return.

Skill 19: Pride

The misleading image of pride has been warped by the advancement of connections. These methods lead us to believe we are far more progressive than reality shows. We credit our successes based upon a false system. What is the price for this satisfaction? We can potentially shame employees, coworkers, partners, or others by the parade of results, numbers, and projections.

A fatal outcome results when we learn to play by the system's rules. We will lose our humanity. The algorithms force control on the masses; they cheat the common man. Our lives become robotic, and we lose the ability to communicate away from a technology driven society. The connections deprive us.

Our lives become a façade; everyone looks at a screen to swipe or click. Society has lost its individuality. Their identities are washed up into the system. Pride envelops us with a dark veil, keeping us trapped in a system of false beliefs. But...this life is a lie.

Part 4: The Physical System

Skill 20: Excess

This skill focuses on living in the present and not the past. An excess of anything develops unhealthy habits. Bad habits interrupt our lives and prevent us from achieving greatness. Negativity not only jeopardizes your health—both mentally and physically—it is a waste of time and energy. Ask yourself, why am I hanging onto a plague when I can swap it with healthy successful practices?

Habits often stem from deeper issues. Stress and boredom are just residual effects. These problems are tough to acknowledge, but change requires commitment. Be honest with yourself.

Individual beliefs bring these practices to the surface without the person's conscious understanding. Stop excessive behavior before developing anxiety-filled compulsions. Connections to worldwide associations are exciting concepts and can be beneficial with proper management.

Data miners are only seeking one thing from the user: tracking habits and designing connections to keep them addicted to clicking. The cycle is perpetual. Information is jammed in every corner of the Internet and comes at you with lightning speed. Excessive overload ensures you stay connected.

The urge to maintain a constant connection with your technological device is in the middle of everyone's life. It is an avoidance practice to ignore real issues. The habit divides your attention and destroys productivity. (Therefore, simplistic advice like, "just stop doing it" rarely works.) The feelings remain dominant, so ignoring them won't diffuse their strength. Here is an instance where going cold turkey could cause additional hardships. If smoking is used to relieve anxiety, you might turn to other unhealthy options to fill the void.

Find an outlet for the stress that has proven effects on your mental and physical well-being. One incredible option is exercise. Small amounts each day can foster positive results.

Skill 21: Desire

Today's society has formed a connection world filled with instant gratification. Desire creates distractions that lure us into the advertising hook created by ever-advancing technology. All aspects of self-discipline have vanished in this connected world. We must become judicious in our efforts to control data mining across the Internet. Uncontrolled Internet usage allows algorithms to reprogram your freedom, turning your choices into robotic lusts that click to fulfill its malevolent cycle. Caution must be taken to control emotional attachment to Internet usage. It's easy to create a false world filled with fantasies that force disconnection from reality. Therefore, you must check your desires, or they will become monsters that haunt your subconscious mind.

Skill 22: Greed

The new technological world has become an intense, selfish existence. It garners wealth through the power of intelligence gathering to collect data. The information is filtered to control actions before you can think any thoughts. A human who spreads wealth to aid others is called a Good Samaritan. But those that claim to help resolve the plight of society through power and fear are self-deceived fools. The plagues are far subtler and more deceptive with the imitation of data collection.

Skill 23: Absence

In this step, we learn how Internet algorithms focus on weakness to gain an advantage in data mining for profit. Imagine each time you click on something, regardless of importance, your action is tracked and documented to draw you into a deeper prison. The data is collected to take control over your life. It's time we learn boundaries when it comes to granting the governorship of our personal identities. So, where do you draw the line and begin to detach from the connections?

Once the decision to withdraw from technology is chosen, the absence will create a void that must be filled with a healthy activity. This step will take mindful diligence.

Daily life can become monotonous, and we function on "autopilot." The less attentive you are to normal rituals, the more gaps open up in your plans to alter behavior. Counteract bad habits with good behaviors to avoid the circumstance and focus on the endgame.

Concentration teaches our brain to react in different ways to certain situations. It can reprogram how you respond to events and stressors. This change allows time to think before you react, breaking the "automatic pilot."

- Be conscious of temptations.
- What triggers lead to unwanted action?
- What sensations in your body or thoughts promote undesired behaviors?

The last precaution to be aware is likely the most relevant section in this step—do not suppress feelings about your habit. Your brain, ironically, will continuously emit impulses if you ignore a desire. The concept is known as "avoidance" (a more familiar term is "procrastination") and will enhance the situation to make matters worse. Conscious acknowledgment will overcome unwanted behavior.

Part 5: The Technology Illusion

Skill 24: Simplicity

Societal pressures have pushed humans beyond the breaking point. We must learn to slow down and remove the complications that program our minds to overdose on Internet usage. Think back to the times when you were disconnected from technology. What happened?

Several things will occur; you either seek refuge in the silence, or the side effects of a craving come forth. We don't realize the damage created from an overdose of technology until it is removed from our lives. The habit of simplicity must be enacted. Severity is an efficient way to yield productivity. When unencumbered with the grips of technology, the chains of innovation are enabled.

Skill 25: Humbleness

The point of this section is understanding the meaning behind humility. We must learn to put the needs of another person ahead of our own. Self-examination allows us to see our true place within society. When we remove pride from the equation and work to seek a humble frame of mind, the rewards of life will be revealed. However, these chains cannot be broken until we become aware of the binding ties we have developed with technology. At that point, we can be free to live a life of humility.

Awareness is important to show what traps hinder our progress, the ties that bind us to this life of technological dependence. Only then will you truly be free to live a life of humility.

Skill 26: Discernment

A true leader is created from within; technology will never fulfill your longings for happiness. The ability to discern a developed moral compass comes with self-examination. It requires time spent in silence, alone with your thoughts, accepting the mistakes of the past and acknowledging the person inside. Once recognition of your shortcomings is realized, progress can be initiated to seek peace with oneself. But it all starts with acceptance of the past.

Part 6: Union with Technology (The Technology Within Yourself)

Skill 27: Stillness

Thoughts travel the universe. When you connect with love, acceptance, and willingness, an aura forms attracting like-minded individuals. The law of attraction. Imagine the impact. Our thought vibration creates an energy field—if you enter a tense and upset room, the emotions spread. Peace develops from within, and humans are attracted to inner joy.

Have you ever noticed how certain people experience misfortune over and over again while others enjoy success? What about the times you are in a foul mood, but find someone upbeat and your emotion changes immediately? We are connected, for better or worse.

If you want to raise your intellectual state of mind, change the way you think. Listed below are a few suggestions:

- Accept responsibility for your life.
- Examine your beliefs.
- Replace negative, limiting ideals with positive aspirations.

The clamor of technology blocks the ability to see the correct path to peace with our existence.

When the silence takes over, we can visualize the problems of our past and move to resolve the issues. The gripping forces of the Internet want you to click rapidly without conscious thought. It's called subliminal advertising (the use of images and sounds to influence consumers' responses without their conscious mind being activated).

Marketers want to control your every move with any Internet usage. It has forced us to believe being busy is good, and while that has its merits, the brain needs time to recoup. When you learn new things, the brain builds links, but they're inefficient, ad-hoc connections. For your brain to develop efficient pathways, it must prune the associations. The cleaning process happens when you sleep.

A full night's rest will leave you thinking clearly. The pruning frees the pathways, leaving room to synthesize new information. In other words, to learn. The power to be successful cannot be found outside with any form of connection, but within. You possess everything you need to be successful on your terms.

Skill 28: Enlightenment

Philosophers believe rational thought is created by enlightened thought patterns. When the ability to consciously make sense of life begins, you have made the first step toward enlightenment. It grants us the capacity to verify facts and apply logic to our findings. However, we sometimes need assistance to see different views on the truth because our vision may be obstructed. These obstacles can be caused from many variables, but in this case, technology is creating the largest hurdle.

The Unhackable mindset process was designed to be a guide in the enlightenment of your mind concerning any form of technology. Realizing these steps makes resolving issues with Internet usage simple. Once the data mining is stopped, reality versus illusion is clear.

Skill 29: Dispassion

Humans have the ability to be completely dispassionate when we can see reality without being influenced by strong emotions.

The mind remains rational and impartial to events in your environment. You learn to master your senses. Technology tracks every move by following every action made on the Internet. The same technology created to free us from boundary restraints has enslaved us. It is time to create a true identity that has been stripped of inaccurate information used to keep us enslaved. The Unhackable mindset teaches how to take your life back, freed from the imprisonment of connection.

Skill 30: Belief

The final skill allows you to stop fighting the issues with connection usage. You will learn to distinguish the difference between what is needed and what is unwanted garbage. You have reached the age of balance. We can surf without our every action being minded. But we must recreate ourselves, enlightened with the accurate knowledge of safe Internet usage. One final practice everyone should observe is a deep-seated belief in yourself to succeed in any area of life you choose.

When men and women are denied freedom, they must become the pioneers of emancipation. To remove the chains of connectivity and loosen the bonds of Internet slavery, they must free themselves or always remain the captive of technology and innovation.

Develop Your Unhackable Mindset!

It is time to become your own security shield! Go to your *Unhackable Workbook* and complete the exercises and questions for Part III. Within the workbook, begin shifting your mindset about how you use technology and identify the first three steps that you will implement to protect your data.

To access your complimentary *Unhackable Workbook*, go to www.GeorgeMansour.com/workbook

PART IV

A PARADIGM SHIFT

Now Is the Time to Take Back Control

13

A DIGITAL TRANSFORMATION

"The Fourth Industrial Revolution is not just about technology or business. It's about society."
—Joe Kaeser

The Internet has become a dangerous place for the normal user. It's a breeding ground for professional hackers, conglomerates seeking to store your data, and businesses marketing an increase in instant gratification purchases. Most of these processes are aimed at financial gain, through devious manipulation. Staying secure under the current conditions is impossible.

Hackers troll hidden areas of the Internet (dark web) for personal data to steal, then sell the information for profit. When a user activates the Unhackable mindset, peace of mind will resume. But the process does not end there. The problem is a societal issue. We must come together as one unit to free ourselves from the Internet manipulation that is destroying the human connection.

Technology's use must alter its focus to the betterment of

humanity, changing one mind at a time. Once the transformation has begun, no company or organization will be more powerful than the people. Are you ready to stand together and overcome the mighty powers of big business promoting technology algorithms that seek to control our online usage?

Businesses must make money to succeed, but at what cost? Entrepreneurship means solving problems while working to change the world for the benefit of humanity. Recent terms for profiting like "Social Entrepreneurship," "Social Enterprises," and "Impact Investing" have been invented to help describe a growing trend. True entrepreneurs focus on a vision for the generations to come. Financial independence can be achieved and a business can generate a solid ROI without endangering others in the process.

Technological innovations have found many new ways to leverage mobile markets, robotics, and crowdsourcing to solve social problems. But we must seek connections that execute a positive effect on humanity. We must remove all lines of demarcation to be victorious and achieve total control over our lives.

The decision must be a requisite if society is to overcome our current status. Technology is a required implement of modern culture and the only way humans will create a more advanced civilization. But the transformation starts from within each user. As I see it, we have two choices: either we fall in the midst of chaos, or we rebuild from the ground up, developing the Unhackable mindset as the foundation. We can truly transform our mindset and create a safe environment for ourselves and future generations.

The Golden Rule

Everyone has heard of The Golden Rule: Do unto others as you would have done unto you. The Golden Rule has been around forever as a conduct of men, but we have missed the meaning and spirit of this universal order. We understand the philosophy as an ethical code of conduct, but we have failed to understand the law upon which it's based. What is the real reason for this kind consideration of others? When you select this rule of conduct as a guide for yourself in your transactions with others, you will be fair and just. The challenge is applying it in our lives, on a daily basis, in everything we do. Is this really happening today?

Philip Zimbardo is an American psychologist best known for his Stanford prison experiment.

> Zimbardo's primary reason for conducting the experiment was to focus on the power of roles, rules, symbols, group identity and situational validation of behavior that generally would repulse ordinary individuals. "I had been conducting research for some years on deindividuation, vandalism and dehumanization that illustrated the ease with which ordinary people could be led to engage in anti-social acts by putting them in situations where they felt anonymous, or they could perceive of others in ways that made them less than human, as enemies or objects," Zimbardo told the Toronto symposium in the summer of 1996.[1]

In his TED talk, "The Psychology of Evil," he shares what turns good people bad. However, he also shares how we can collectively and individually make positive changes:

> There are seven social processes that grease "the slippery slope of evil":

> Mindlessly taking the first small step

> Dehumanization of others

> Deindividuation of self (anonymity)

Diffusion of personal responsibility

Blind obedience to authority

Uncritical conformity to group norms

Passive tolerance of evil through inaction or indifference

So you need a paradigm shift in all of these areas. The shift is away from the medical model that focuses only on the individual. The shift is toward a public health model that recognizes situational and systemic vectors of disease. Bullying is a disease. Prejudice is a disease. Violence is a disease. Since the Inquisition, we've been dealing with problems at the individual level. It doesn't work. Aleksandr Solzhenitsyn says, "The line between good and evil cuts through the heart of every human being." That means that line is not out there. That's a decision that you have to make, a personal thing.[2]

Unfortunately, most of the advances in the technology field are merely for the advancement of self, rather than others. If we were thinking about what is best for others, we would not all be stuck in this rat race. Society can function as one unit in peace, if we learn to work as a team.

As a security expert, I'm going against traditional norms. I understand the importance of doing what's right for the good of others. Life is not all about financial gain.

Business has become about ROI, marketing, collecting data, and exposing our lives to the public. Others are able to utilize that data for malicious acts. Marketing has become duplicitous and illusory, blinding us from what is truly happening. The business world is not about bettering you but bettering the business. The transformation of the information highway has developed a dangerous situation for all users. It's time to return to "of the people, for the people, by the people."

The Internet is now over four decades old. A survey of its evolution from a military experiment conducted in the context of the Cold War to a General Purpose Technology illustrates the extent to which the network was shaped, not just by the intrinsic affordances of its underpinning technologies, but also by political, ideological, social, and economic factors.[3]

How Do We Fix Our Broken System?

While most corporations are focused on their return on investment, some businesses are using technology to better our world. Four Thieves Vinegar (www.fourthievesvinegar.org) continues their mission to make medicine free for everyone. "We have designed an open-source automated lab reactor, which can be built with off-the-shelf parts, and can be set to synthesize different medications. It will save hundreds of thousands of lives."[4] One priority in life should be finding ways to contribute selflessly for the betterment of humanity through technology. The human race works better as a unified team.

Specialization through Collaboration

When a society is focused on financial gain, or the improvement of oneself, humanity suffers. Businesses are not focused on how they can collaborate to plan for the future. Collectively, we can improve life for everyone on the planet, rather than contribute to its destruction.

The problem was created by everyone, not one person or company, therefore it must be resolved as a team. Empowered users can specialize in the dynamics of cybersecurity, therefore no one person is left outside the schooling pod. The group moves in fluid movement to create a completely secure system. Once you have mastered the Unhackable mindset thought process, free from the boundaries created by society, you'll clearly identify the correct path.

Accountability

Large conglomerates that control the vast majority of businesses must be held accountable for any action that affects clients in a detrimental fashion. Accountability can come from shareholders, consumers, employees, our institutions, and even shareholders. "Once uncritically hailed for their innovation and economic success, Silicon Valley companies are under fire from all sides, facing calls to take more responsibility for their role in everything from election meddling and hate speech to physical health and addiction."[5] One investment firm is holding businesses accountable by withholding investments if they are selfishly motivated by profits, rather than contributing to the greater good. According to *The New York Times*[6]:

> Laurence D. Fink, founder and chief executive of the investment firm BlackRock, is going to inform business leaders that their companies need to do more than make profits—they need to contribute to society as well if they want to receive the support of BlackRock.

Fink isn't the only one putting pressure on companies to think about others and make changes. According to the above article,

> ...even activist investors are taking up social causes. Jana Partners and Calstrs, the huge California retirement system that manages the pensions of the state's public school teachers, wrote a letter to Apple last week demanding that it focus more on the detrimental effects its products may have on children...."In the case of Apple," Jana wrote, "we believe the long-term health of its youngest customers and the health of society, our economy, and the company itself, are inextricably linked."

Facebook is also beginning to take accountability for its

contribution to the negative, disconnected experience users have been having. It has started to make improvements by changing the content in its News Feed. Zuckerberg stated, "We feel a responsibility to make sure our services aren't just fun to use, but also good for people's well-being ...the research shows that when we use social media to connect with people we care about, it can be good for our well-being. We can feel more connected and less lonely, and that correlates with long term measures of happiness and health. On the other hand, passively reading articles or watching videos—even if they're entertaining or informative—may not be as good."[7]

Although the theory looks good on the surface, we must continue to voice our concerns until the companies are held accountable for their actions. Our collective voices can rise to the top and demand change.

Change in Business

The way people conducted business changed forever with the invention of the Internet. It opened doors for individuals around the world to communicate at lightning speeds. Innovation brought financial independence to a level unmatched by business norms. But the connection also brought disruptive forces like the cloud, social media, and data analytics to reinvent the way people interact and engage with each other and businesses. The ability to compile data generated an algorithm-driven market to help users make more informed decisions. On the surface, the idea sounds appealing, but when greed overcame human connection, it corrupted the once innocent pathway and began the corruptive unscrupulous data mining found on the Internet today.

The data collection resources combined with behavioral attributes are mined without the knowledge of the normal Internet user. Connections create an entangled web of raw data without visible boundaries to maintain a secure system. In this environment, data is exposed to professional hackers waiting to steal your digital identity.

The World Wide Web was named for its resemblance to a spider web: every strand intertwines with the others to represent a stable connection. Hyperlinks are what make the World Wide Web function. The links allow the user to locate information quickly, but we are losing control of the situation rapidly. Our thoughts are being controlled by algorithms meant to trap us in an uncompromising spider web.

Innovation does not always come with a textbook to teach us about regulations, nor is the user license agreement shared willingly. The responsibility falls upon the user to act within the guidelines, set in place with the Unhackable mindset security defense.

Otherwise, users fall prey to malicious attacks that steal precious data for financial gain. Technology generates a stream of simultaneous marketing blasts to force uncontrolled clicking without thinking about the risks. The result is a breach of privacy.

As a business leader in the digital revolution, think carefully about the interactions your company has with its customers. Research the apps and other technology that may invade their privacy. Safer connections start with calculated business practices. Business owners must also create a healthier relationship with technology for employees. Initiate useful tools rather than daunting burdens. Unhackable mindset protocols must be activated to instill confidence within every employee. Owners must have training classes and clear system regulations to possess a secure defensive environment.

Digital transformation comes from changing not only the way we deal with technology but personal offline habits. Tech culture can be altered by encouraging employees to disconnect at the end of the day, or during vacation. In his TedTalk, Adam Alter provides two examples of companies who are embracing change by incorporating disconnecting rules into their policies.

A Dutch design firm raises the desks to the ceiling every day at 6 pm and transforms the workplace into a yoga studio or dance club.

"At Daimler, the German car company, they've got another great strategy. When an employee goes on vacation, instead of saying, 'The individual is on vacation, your call will be returned at a later date.' They say instead, 'This person's on vacation, so we've deleted your email. You can resend an email in a couple of weeks, or email another person in the meantime.' Employees actually feel as though they get a break from work."[8] The choice will not only change the way we view technology, but minimize stress and reduce our dependence on technology. Powerful leaders are created through the initiation of change.

Education

Organizational leadership roles in academic settings must change to accommodate the ever-changing needs of Internet usage.

A common myth surfaces to tell us growth is the answer to resolving problems, however, evolution alone will not avail. The issues facing society are more than economic stamina; it's about raising the human condition.

Society is being driven by data analytics for all-inclusive development. It's called behavioral economics. The intention began with leveraging growth through the benefits of monetary effects. We must add an objective measure to explain the term inclusively. The increased progress encompasses three modules:

- The sharing of users in troublesome businesses and growing markets, particularly where data mining services exist, will reveal new competence and redirect outdated jobs.
- Generate engagement openings by the free movement of digital information to sectors of society that suffer in

monetary distress; the commercial influx brings benefit to those markets.

- Sensible portion of taxpayer funds for the enhancement of society will help the well-being of everyone, minus unprincipled program-making activity.

The three mainstays are only verified when the majority of the population contributes to the advancement of society. We must educate to improve the well-being of society while simultaneously creating jobs. The empowering factor is technology with its ability to connect with people all around the world.

Advanced processing allows us to collect huge amounts of data while minimizing costs to build the complex copy to ask the unimaginable questions. Analytics bring reality to the wide-ranging evolution facing the world today.

Data mining can be used to enrich the public sector rather than fail and fix it. Instead, we have the information to predict and plan, or better yet, prevent. In other words, the damage cannot be repaired with a band-aid; the loopholes must be closed, allowing a safe cybersecurity environment free from all fraud and risks as you develop the Unhackable mindset. When analytics are used to predict security breaches rather than public duplicity, everyone benefits.

In conclusion, data analytics could transform society to trigger effective programs that promote financial transformation, improving the quality of life for every person on the planet. Prevention interactions have proven valuable in the production of a healthy society.

14

SECURE THE WORLD

"Every generation needs a new revolution."
—Thomas Jefferson

A Future of Freedom

As a society, we must regain control of our innovations. Humans have handed dominion over to technology, instead of remaining inventors.

We can create a universal environment where all data on the Internet is treated equally; neutrality becomes the norm. A situation where everyone works together for prosperity, a tool to advance humanity without the fear of data theft. To change the future, we need to look to the past. How did we get here? "Who Invented the Internet" shares a brief overview of the history of the Internet:

As you might expect for a technology so expansive and ever-changing, it is impossible to credit the invention of the Internet to a single person. The Internet was the work of dozens of pioneering scientists, programmers and engineers who each developed new features and technologies that eventually merged to become the "information superhighway" we know today.

Long before the technology existed to actually build the Internet, many scientists had already anticipated the existence of worldwide networks of information. Nikola Tesla toyed with the idea of a "world wireless system" in the early 1900s, and visionary thinkers like Paul Otlet and Vannevar Bush conceived of mechanized, searchable storage systems of books and media in the 1930s and 1940s. Still, the first practical schematics for the Internet would not arrive until the early 1960s, when MIT's J.C.R. Licklider popularized the idea of an "Intergalactic Network" of computers. Shortly thereafter, computer scientists developed the concept of "packet switching," a method for effectively transmitting electronic data that would later become one of the major building blocks of the Internet.

The first workable prototype of the Internet came in the late 1960s with the creation of ARPANET, or the Advanced Research Projects Agency Network. Originally funded by the U.S. Department of Defense, ARPANET used packet switching to allow multiple computers to communicate on a single network. The technology continued to grow in the 1970s after scientists Robert Kahn and Vinton Cerf developed Transmission Control Protocol and Internet Protocol, or TCP/IP, a communications model that set standards for how data could be transmitted between multiple networks. ARPANET adopted TCP/IP on January 1, 1983, and from there researchers began to assemble the "network of networks" that became the modern Internet. The online world then took on a more recognizable form in 1990, when computer scientist Tim Berners-Lee invented the World Wide Web. While it's often confused with the Internet itself, the web is actually just the most common means of accessing data online in the form of websites and hyperlinks. The web helped popularize the Internet among the public, and served as a crucial step in developing the vast trove of information that most of us now access on a daily basis. [1]

Innovation is at a crucial junction. We will either demolish society due to our actions or recreate humanity with a better version of technology. The Internet relies on each user to take responsibility for their actions. Change comes from studying the past, living in the present, and creating a better version of society.

Once we regain control of our lives, harmony will replace the plagues attacking our sanity. Internet dependence has become a serious issue with many people worldwide. Technology was meant to connect humans around the world, but instead the opposite effect has continued to put a stranglehold on users. We must become the generation of change and neutralization, take back control of our innovations and seek a more harmonious future. The only way to secure technology is to neutralize the instant gratification impulses of Internet use.

Together we can form a movement to repair the failures in technology, thereby creating the transformational leadership of future generations. Instinct-based control over current Internet users will return to human connections, and free our world from non-thinking connection followers.

Your Part in the Paradigm Shift

The world needs inspirational leaders to help guide us through this uncertain environment with all of the connected devices and scattered data in our lives. The future will be filled with even more devices, and we cannot let them take over. Cisco's Complete VNI report forecasts global IP traffic growth for mobile and fixed networks. By 2022, there will be 4.8 billion global Internet users and 28.5 billion networked devices and connections.[2]

Attackers who use threats find joy in our weakness, because that's all they have. Hackers don't know what true connection is. "Maslow's hierarchy of needs is a theory in psychology comprising a five-tier model of human needs, often depicted as hierarchical levels within a triangle pyramid. From the bottom of the hierarchy upwards, the needs are: physiological, safety, love and belonging, esteem and self-actualization. Needs lower down in the hierarchy must be satisfied before individuals can attend to needs higher up."[3]

Once we have food and shelter, but before we can seek self-actualization, we must develop a nurturing environment. If someone lacks these three essential elements, we cannot perform, innovate, or feel emotionally engaged. So, let's destroy all threats from the ground up and build a solid foundation worthy of defense—for the people, by the people.

Tell yourself today, "I will not bow down or be broken. I will take another step and overcome all the problems I face today."

That is the true power we all have in us that defines what true humanity should be about. True power is achieved when two or more come together with similar values aligned, then work toward a common goal. The process is called the true mastermind effect. Together, we shall overcome this as a united front equipped and ready to neutralize with our own Unhackable mindset.

Develop Your Unhackable Mindset!

You now have the knowledge to begin your transformation and develop an Unhackable mindset. But it isn't just about building a secure network; the alteration is seeking to educate and support fellow humans to find peace with their Internet connections.

Go to your *Unhackable Workbook* and complete the exercises and questions for Part VI. Within the workbook, identify ways you can begin to share your newfound knowledge with others and reflect on the next training steps you need to take to become your own security shield.

If you have not yet accessed your complimentary *Unhackable Workbook*, go to www.GeorgeMansour.com/workbook

AFTERWORD

As you process the information in this book and the reality that has been revealed, how are you going to change your current mindset? It is time for a paradigm shift. Are you going to be part of the change? It is clear that on our current path nothing is secure or private. The issue will continue to escalate unless we take action immediately. But it doesn't have to be this way. We can change the future. We must work as a team to slow the advancement of technology while cleaning up the mess in the process.

Unless we engage in a major transformation, we are going to lose our humanity. We need to wake up; restore, reconstruct, and rebuild digital technology to serve and connect us to a truer self. If we do not question our digital security, then how can we ever be safe?

Join the Unhackable paradigm shift at www.GeorgeMansour.com. You will have access to up-to-date information about the current technological dangers and the preventative steps necessary to stay secure. Additionally, you will become a master craftsman who can help others build their own cybersecurity defense around their digital data.

ABOUT THE AUTHOR

George Mansour has a unique and simplified approach to cybersecurity. As a trusted IT specialist for more than twenty years, he helps clients overcome the debilitating effects of cybersecurity issues. His distinctive methods have allowed him to emerge as an industry leader for individuals and businesses seeking to find a more secure system.

George understands the imminent threat facing anyone connected online. His goal is to empower end-users with the tools to secure their digital assets using proper protocols. He seeks to enhance the validity of his client's lives, by teaching them the basics of online security. The systems, products, and services he uses are designed to prevent a breach before it happens. Preventative methods are always preferred. It is imperative that users become proactive, because cognitive distortion has taken over our senses. We need a compass to navigate the Internet in real-time, and we need to develop a license agreement with ourselves that isolates our unique, sensitive data.

George received a Bachelor of Science in Business Administration (BSBA) degree with a major in Computer Information Systems (CIS) from Suffolk University, along with a Microsoft Certified Professional (MCP) certification. Besides his academic accomplishments, he is the founder and owner of CEHIT, INC., an acronym for Computer Engineering Hardware Information Technology—an Information Technology company that has helped manage complex IT environments for over 15 years. CEHIT, INC. has been helping thousands of consumers and businesses (small, medium, and large) across the globe. George Mansour focuses on numerous vertical markets and organizations in the healthcare, law, insurance, manufacturing industries, etc. He has affiliations with partners and value-added reseller programs. Cybersecurity is a shared responsibility.

George's message is that technological interaction affects everyone. We are interwoven over the Internet as one, but individually unique. The Unhackable mindset creates the conceptual model to enact a multi-layer digital union between technology and human interaction. The insecurities and impurities that threaten us online with every new connection must be controlled through a collaborative approach that will suppress threats to our sensitive data. He says it's time to reclaim our digital lives and our digital freedom.

Hearing George's message and getting his unmatched advice and step-by-step approach to cybersecurity is something everyone should immediately add to their to-do list! Don't miss out on your opportunity to work with George and be a part of this tech paradigm shift. Visit www.GeorgeMansour.com to learn more about becoming Unhackable.

NOTES

1. Awareness Is Everything

1. Kemp, Simon. "The Incredible Growth of the Internet over the past Five Years - Explained in Detail." *The Next Web*. 04 Mar. 2019. Web.
2. Interactive, THINK. "BY 2020 4 BILLION PEOPLE." LinkedIn SlideShare. 17 Apr. 2014. Web.
3. "Computer Security." *Wikipedia*, Wikimedia Foundation, 22 June 2019. Web.
4. "You Don't Know Hack: Public Struggles with Cyber Security Concepts." The Security Ledger. 23 Mar. 2017. Web.

2. Our Changing Relationship with Technology

1. Alfred, Randy. "April 4, 1975: Bill Gates, Paul Allen Form a Little Partnership." Wired. Conde Nast, 03 June 2017. Web.
2. Curtis, Sophie. "Bill Gates: A History at Microsoft." The Telegraph. Telegraph Media Group, 04 Feb. 2014. Web.
3. Weller, Chris. "Bill Gates Just Bought 25,000 Acres in Arizona to Build a New 'smart City'." Business Insider. Business Insider, 13 Nov. 2017. Web.
4. Rettner, Rachael. "Apple Obsession: The Science of IPad Fanaticism." LiveScience. Purch, 02 May 2010. Web.
5. Weinberger, Matt. "The World of Technology Is Changing and the IPad Is Getting Caught in the Middle." Business Insider. Business Insider, 01 Feb. 2017. Web.
6. Josh Lowensohn April 25, 2012 4:08 PM PDT @Josh. "Kaspersky: Mac Security Is '10 Years behind Microsoft'" CNET. 25 Apr. 2012. Web.
7. Smith, Dave. "The Steve Jobs Guide to Manipulating People and Getting What You Want." Business Insider. Business Insider, 05 Oct. 2016. Web.
8. Wolverton, Troy. "Apple Confirmed a Longtime Conspiracy Theory - and Gave Regular Customers a Big Reason to Distrust It." Business Insider. Business Insider, 20 Dec. 2017. Web.
9. Wolverton, Troy. "Apple Confirmed a Longtime Conspiracy Theory - and Gave Regular Customers a Big Reason to Distrust It." Business Insider. Business Insider, 20 Dec. 2017. Web.

3. Modern Data Gathering

1. "Cybersecurity Isn't Just an IT Problem-It's Also a Marketing Problem." EMarketer. EMarketer, 12 Dec. 2017. Web.

2. Duhigg, Charles. "How Companies Learn Your Secrets." *The New York Times.* The New York Times, 18 Feb. 2012. Web.

3. Hill, Kashmir. "Target Isn't Just Predicting Pregnancies: 'Expect More' Savvy Data-Mining Tricks." *Forbes.* Forbes Magazine, 28 Feb. 2012. Web.

4. Madrigal, Alexis C. "I'm Being Followed: How Google and 104 Other CompaniesAre Tracking Me on the Web." *The Atlantic.* Atlantic Media Company, 29 Feb. 2012. Web.

5. Borzykowski, Bryan, and Special To CNBC.com. "The Chilling Truth about Cybercriminals—from a Paid Hacker." *CNBC.* CNBC, 26 July 2016. Web.

6. Kharpal, Arjun. "Hackers Are Selling Your Data...for Only $1." *CNBC.* CNBC, 24 Sept. 2015. Web.

7. "What Is Data Breach? - Definition from WhatIs.com." *SearchSecurity,* 24 Sept. 2015. Web.

8. Progress, Work In. "The Psychology And Philosophy Of Branding, Marketing, Needs, And Actions." *Forbes.* Forbes Magazine, 05 Mar. 2014. Web.

9. Lock, S. "McDonald's: Ad Spend 2019." *Statista.* 27 Feb. 2020. Web.

10. Grauer, Yael. "What Are 'Data Brokers,' and Why Are They Scooping Up Information About You?" *Motherboard.* VICE. 27 Mar. 2018. Web.

11. "Time Well Spent." *Time Well Spent.* 18 May 2016. Web.

12. "How Technology Is Hijacking Your Mind-from a Former Insider." *Thrive Global.* Thrive Global, 18 May 2016. Web.

13. *Dopamine.* 18 May 2016. Web.

14. "We're the (Hacker) Neuroscientists @ Dopamine Labs." *Dopamine.* 18 May 2016. Web.

15. "Take Control." *Time Well Spent.* 18 May 2016. Web.

4. Cybersecurity Issues

1. "Content Data." The IT Law Wiki. 24 Feb. 2017. Web.

2. https://pdf.ic3.gov/2015_IC3Report.pdf

3. Spitzer, Julie. "Atlanta's Ransomware Attack May Cost the City $17M." *Becker's Hospital Review.* 24 Feb. 2017. Web.

4. "Hackers Targeted DC Police Cams Days Before Inauguration." *TechNewsWorld.com.* 24 Feb. 2017. Web.

5. "Microsoft Seeks Global Cybersecurity Accord." *TechNewsWorld.com.* 24 Feb. 2017. Web.

6. "2012 LinkedIn Hack." *Wikipedia,* Wikimedia Foundation, 18 Mar. 2019.

7. "Flashlight Apps Are Spying on Users Android, IOS, Windows Phone Smartphones, Is Yours on the List?" *Tech Times.* 26 Oct. 2014. Web.

8. Kimmorley, Sarah. "Elon Musk's Former Tech Guru Says What Hackers Are Doing Now Is 'Freaking Me Out'" *Business Insider.* Business Insider, 24 Feb. 2017. Web.

9. "'We Live in a New World of Sophisticated Hacking & Cryptojacking – McAfee to RT." *RT International.* 12 Mar. 2018. Web.

10. Boden, Sarah. "Be Wary Of Tech Support Scams, 'Companies Like Microsoft Are Not Actually Going To Call You'." *90.5 WESA.* 12 Mar. 2018. Web.

11. "Hackers Use Facebook Quizzes to Steal Personal Info." *NBCNews.com*. NBCUniversal News Group, 12 Mar. 2018. Web.

12. CBS Boston. "Police Warn Of 'Can You Hear Me?' Phone Scam." *CBS Boston*. CBS Boston, 30 Mar. 2017. Web.

13. Dickler, Jessica. "Forget a Data Breach, Consumers Give Away Their Personal Information on Social Media." *CNBC*. CNBC, 11 Mar. 2018. Web.

14. Suciu, Peter. "Microsoft Seeks Global Cybersecurity Accord." *TechNewsWorld.com*. 19 Apr. 2018. Web.

15. Lord, Debbie. "Apple Issues Urgent IPhone IOS Upgrade; How to Protect Your Phone." *Daytondailynews*. Cox Media Group National Content Desk, 15 Sept. 2016. Web.

16. Cox Media Group National Content Desk. "Apple Issues Urgent IPhone IOS Upgrade; How to Protect Your Phone." *WFXT*. 26 Aug. 2016. Web.

17. "Identity Theft." *Bureau of Justice Statistics (BJS)*. 26 Aug. 2016. Web.

18. "Victims of Identity Theft, 2014." *Bureau of Justice Statistics (BJS)*. 26 Aug. 2016. Web.

19. "Silent Epidemic: Data Theft Has Become a Public Health Crisis | Digital Guardian." *The Security Ledger*. 27 Feb. 2017. Web.

20. Szoldra, Paul. "More than 86% of the World's IPhones Can Still Be Hacked with Just a Text." Business Insider. Business Insider, 29 Aug. 2016. Web.

21. "CIA Hacks TVs, Phones All over the World, WikiLeaks Claims." *CNNMoney*. Cable News Network. 4 Sept. 2016. Web.

22. Pagliery, Jose.. "NBC to Air Edward Snowden Interview." *CNNMoney*. Cable News Network. 22 May 2014. Web.

23. Pagliery, Jose. "How the NSA Can 'Turn On' Your Cell Phone Remotely." *CNNMoney*. Cable News Network. 6 June 2014. Web.

24. Zuckerbrod, Nancy. "Report: Govt. Web Sites Invade Privacy." *ABC News*. ABC News Network. 7 Jan. 2006. Web.

25. Daileda, Colin. "WikiLeaks Document Dump Alleges the CIA Can Hack Almost Everything." *Mashable*. Mashable, 07 Mar. 2017. Web.

26. "How to Find out Everything That Google Knows about You." *CNBC*. CNBC, 06 Dec. 2017. Web.

27. Griffin, Andrew. "Google Voice Search Records and Keeps Conversations People Have around Their Phones–but the Files Can Be Deleted." *The Independent*. Independent Digital News and Media, 01 June 2016. Web.

28. Epstein, Robert. "Google's Gotcha." *U.S. News & World Report*. U.S. News & World Report, 10 May 2013. Web.

29. "Is Google Maps a Threat to Privacy?" *Our Everyday Life*. 10 May 2013. Web.

30. Fiorella, Sam. "The Insidiousness of Facebook Messenger's Android Mobile App Permissions (Updated)." *The Huffington Post*. TheHuffingtonPost.com, 07 Dec. 2017. Web.

31. "Facebook Wants to Listen to Your Phone Calls." *Infowars*. 27 Nov. 2013. Web.

32. Gibbs, Samuel. "Your Facebook Messenger App Is about to Be Filled with Ads." *The Guardian*. Guardian News and Media, 12 July 2017. Web.

33. Hu, Elise. "Facebook Manipulates Our Moods For Science And Commerce: A Roundup." NPR. NPR, 30 June 2014. Web.

34. Granville, Kevin. "How Cambridge Analytica Harvested Facebook Data,

Triggering a New Outcry." *The New York Times*. The New York Times, 19 Mar. 2018. Web.

35. Steel, Emily, and Geoffrey A. Fowler. "Facebook in Privacy Breach." *The Wall Street Journal*. Dow Jones & Company, 18 Oct. 2010. Web.

36. Haselton, Todd. "How to Download a Copy of Everything Facebook Knows about You." *CNBC*, CNBC, 27 Mar. 2018. Web.

37. Haselton, Todd. "How to Download a Copy of Everything Facebook Knows about You." *CNBC*. CNBC, 24 Mar. 2018. Web.

38. John Johnson Newser. "Girl Uses Sleeping Mom's Thumbprint to Buy $250 in Pokemon Toys." *USA Today*. Gannett Satellite Information Network, 29 Dec. 2016. Web.

39. King, Mark. "Parents Told to Beware Children Running up Huge Bills on IPad and IPhone Game Apps." *The Observer*. Guardian News and Media, 12 Jan. 2013. Web.

5. Cyber Dangers of the Future

1. Deutsche Welle. "Saudi Arabia Grants Citizenship to Robot Sophia | News | DW | 28.10.2017." *DW.COM*. Web.

2. Jones, Rhett. "Roomba's Next Big Step Is Selling Maps of Your Home to the Highest Bidder." *Gizmodo*. Gizmodo.com, 24 July 2017. Web.

3. Jones, Rhett. "Roomba's Next Big Step Is Selling Maps of Your Home to the Highest Bidder." *Gizmodo*. Gizmodo.com, 24 July 2017. Web.

4. Stella, Rick. "Facebook's Mark Zuckerberg Pulls Back the Veil on the Jarvis AI System He Created." *Digital Trends*. 20 Dec. 2016. Web.

5. Condliffe, Jamie. "This Year, We Learned to Love AI Assistants in Our Homes." *MIT Technology Review*. MIT Technology Review, 21 Dec. 2016. Web.

6. Terdiman, Daniel. "At Home With Mark Zuckerberg And Jarvis, The AI Assistant He Built For His Family | Fast Company | The Future Of Business." *Fast Company*. Fast Company, 10 Mar. 2017. Web.

7. Hern, Alex. "Google Accused of Spreading Fake News." *The Guardian*. Guardian News and Media, 06 Mar. 2017. Web.

8. "IoT Mobile FINAL." Interactive Content Marketing Examples Created With Ceros. 06 Mar. 2017. Web.

9. Kadvany, Elena. "Los Altos Planning Commission Chair Arrested for Tesla DUI." Palo Alto Online - Lasting Memories - Virginia Belle Bandura's Memorial. 06 Dec. 2018. Web.

10. Cava, Marco Della. "Warning, Drivers: Your Car Can Cruise on Its Own, but You're Still Responsible." *USA Today*. Gannett Satellite Information Network, 29 May 2018. Web.

11. "The Cybersecurity Risk of Self-driving Cars." *Phys.org - News and Articles on Science and Technology*. 16 Feb. 2017. Web.

12. Mlot, Stephanie. "These Internet-Connected Toys Provide Joy (and Surveillance)." *PCMAG*. PCMAG.COM, 08 Dec. 2016. Web.

13. Picchi, Aimee. "Your Kids' Toys Could Be Spying on Your Family." *CBS News*. CBS Interactive, 07 Dec. 2016. Web.

14. "Instant Access to Gartner Research on Tech Trends and Emerging Technology." *Gartner*. Web.
15. "In Ten Years, Robots Could Replace More than 4 Million Workers." *Futurism*. Futurism, 20 Sept. 2017. Web.
16. Primack, Dan. "These Are the Jobs That Artificial Intelligence Will Eliminate First." *Fortune.com*. Fortune, 12 July 2016. Web.
17. "Bitcoin-Open Source P2P Money." *Bitcoin-Open Source P2P Money*. Web.
18. Astor, Maggie. "Microchip Implants for Employees? One Company Says Yes." *The New York Times*. The New York Times, 25 July 2017. Web.
19. Nikolov, Nikolay. "These Tattoos Conduct Electricity, Turning You into a Very Basic Cyborg." *Mashable*. Mashable, 06 Mar. 2017. Web.
20. Farahany, Nita. "Transcript of "When Technology Can Read Minds, How Will We Protect Our Privacy?"" Ted. Ted, 6 Dec. 2018. Web.
21. Newman, Lily Hay. "What We Know About Friday's Massive East Coast Internet Outage." *Wired*. Conde Nast, 21 Oct. 2016. Web.
22. Zetter, Kim. "An Unprecedented Look at Stuxnet, the World's First Digital Weapon." Wired. Conde Nast, 03 June 2017. Web.
23. Fruhlinger, Josh. "What Is Stuxnet, Who Created It and How Does It Work?" CSO Online. CSO, 22 Aug. 2017. Web.
24. Phil Stewart, Diane Bartz, Jim Wolf, and Jeff Mason. "Special Report: Government in Cyber Fight but Can't Keep up." *Reuters*. Thomson Reuters, 16 June 2011. Web.
25. "FBI's Comey: 'You're Stuck with Me for Another 6 Years'" *ABC News*. ABC News Network. 8 Mar. 2017. Web.
26. Cat Loves CNN. "War In Space-The Next Battlefield-CNN." *YouTube*, YouTube, 29 Nov. 2016. Web.

6. A Greater Threat

1. "Captivated Documentary–Finding Freedom in a Media Captive Culture." *Captivated Documentary*. Web.
2. "IoT-enabled Devices to Top 40 Billion by 2020, Say Researchers." ComputerWeekly.com. 25 Sept. 2014. Web.
3. Summers, Juana. "Kids And Screen Time: What Does The Research Say?" *NPR*. NPR, 28 Aug. 2014. Web.
4. Knapton, Sarah. "Using IPads to Pacify Children May Harm Their Development, Say Scientists." *The Telegraph*. Telegraph Media Group, 01 Feb. 2015. Web.
5. Weller, Chris. "Bill Gates and Steve Jobs Raised Their Kids Tech-free - and It Should've Been a Red Flag." *Business Insider*. Business Insider, 10 Jan. 2018. Web.
6. Bilton, Nick. "Steve Jobs Was a Low-Tech Parent." *The New York Times*. The New York Times, 10 Sept. 2014. Web.
7. CBS News. "Silicon Valley Exec on Regulating Facebook and Why He Keeps His Kids Away from Screens." *CBS News*. CBS Interactive, 08 May 2018. Web.
8. Gates, Melinda. "Perspective | Melinda Gates: I Spent My Career in

Technology. I Wasn't Prepared for Its Effect on My Kids." The Washington Post. WP Company, 24 Aug. 2017. Web.

9. "HOME | Screenagers Movie." *SCREENAGERS*. Web.

10. *SCREENAGERS*, www.screenagersmovie.com/family-contract.

11. O'Donnell, Jennifer. "How to Make a Cell Phone Contract With Your Child That Works." *Verywell Family*, Verywell Family, 27 Aug. 2018. Web.

12. "Captivated Documentary—Finding Freedom in a Media Captive Culture." *Captivated Documentary*. Web.

13. "Captivated Documentary—Finding Freedom in a Media Captive Culture." *Captivated Documentary*. Web.

14. "How Technology Is Controlling Our Relationships." *Odyssey*. 05 July 2016. Web.

15. "Negative Effects Of Social Media On Your Health." *Negative Effects Of Social Media On Your Health*. Web.

16. https://pdf.ic3.gov/2015_IC3Report.pdf

17. Twenge, Jean M. "Have Smartphones Destroyed a Generation?" *The Atlantic*. Atlantic Media Company, 19 Mar. 2018. Web.

18. Alter, Adam. "Why Our Screens Make Us Less Happy." *TED: Ideas worth Spreading*. 14 Jul. 2017. Web.

19. Susarla, Anjana. "How Artificial Intelligence Can Detect–and Create–Fake News." *Big Think*. Big Think, 05 May 2018. Web.

20. Granville, Kevin. "How Cambridge Analytica Harvested Facebook Data, Triggering a New Outcry." The New York Times. The New York Times, 19 Mar. 2018. Web.

21. Pearle, Lauren. "Cambridge Analytica Accused of Violating US Election Laws in New Legal Action." *ABC News*. ABC News Network, 26 Mar. 2018. Web.

22. Stanfordbusiness. "Chamath Palihapitiya, Founder and CEO Social Capital, on Money as an Instrument of Change." *YouTube*. YouTube, 13 Nov. 2017. Web.

7. Train Your Brain to be the Security Shield

1. Cat Loves Cnn. "War In Space: The Next Battlefield—CNN." *YouTube*. YouTube, 29 Nov. 2016. Web.

2. Stanfordbusiness. "Chamath Palihapitiya, Founder and CEO Social Capital, on Money as an Instrument of Change." *YouTube*. YouTube, 13 Nov. 2017. Web.

9. Balance the Old and the New

1. "PERSUASIVE TECHNOLOGY LAB." *WHAT IS CAPTOLOGY? – Persuasive Tech*, captology.stanford.edu/about/what-is-captology.html.

2. "PERSUASIVE   TECHNOLOGY   LAB."   *Persuasive   Tech*, captology.stanford.edu/.

3. Chatfield, Tom. "What Does It Mean to Be Human in the Age of Technology?" *The Guardian*, Guardian News and Media, 20 Jan. 2016. Web.

4. "Cognitive Distortion." *Wikipedia*, Wikimedia Foundation, 12 May 2019. Web.

10. Reinventing Business Technology & Security

1. Hiner, Jason. "Is Perimeter Security Dead and Is Protecting the Data All That Matters?" *TechRepublic*. 24 Aug. 2007. Web.

2. "Influencers." *The Free Dictionary*. Farlex. Web.

3. "Online Trust Alliance Reports Doubling of Cyber Incidents in 2017." *Online Trust Alliance*. 25 Jan. 2018. Web.

4. Gittleson, Kim. "Can a Company Live Forever?" *BBC News*. BBC, 19 Jan. 2012. Web.

5. "Cybersecurity: Employees Are the First Line of Defense." *ThinkHR Human Powered*. 14 Nov. 2018. Web.

6. "Online Trust Alliance Reports Doubling of Cyber Incidents in 2017." *Online Trust Alliance*. 25 Jan. 2018. Web.

7. Mathison, David. "CDO, CDO Club, CDO Summit, Chief Digital Officer, Chief Digital Officer Club, Chief Digital Officer Summit, Chief Data Officer, CIO, Chief Information Officer, CMO, Chief Marketing Officer, Chief Marketing Technologist, CTO, Chief Technology Officer, Chief Analytics Officer, Chief Information Security Officer, CISO, David Mathison." *CDO Summit*. Web.

8. "Gartner Estimates That 90 Percent of Large Organizations Will Have a Chief Data Officer by 2019." *Gartner*. 27 Jan. 2016. Web.

9. Dignan, Larry. "Gartner: 'Every Budget Is an IT Budget'" *ZDNet*. ZDNet, 04 Dec. 2015. Web.

10. Auvik Networks. "Threat or Gold Mine? What Lies Ahead for Managed Services." *Auvik Networks*. 08 Nov. 2016. Web.

11. "Data Security Procedures, Computer System Security Requirements." *South Puget Sound Community College*. 14 Aug. 2018. Web.

12. Alsinawi, Baan. "TalaTek Incident Response Services." TalaTek, LLC, talatek.com/risk-management-services/incident-response/.

13. Hiner, Jason. "Is Perimeter Security Dead and Is Protecting the Data All That Matters?" *TechRepublic*. 24 Aug. 2007. Web.

11. Steps to Securing Your Life and Business

1. "Google Ads." *Wikipedia*. Wikimedia Foundation, 11 Feb. 2020. Web.

13. A Digital Transformation

1. The Stanford Prison Experiment: Still Powerful after All These Years (1/97). 8 Jan. 1997. Web.
2. TED. "The Psychology of Evil | Philip Zimbardo." *YouTube*. YouTube, 23 Sept. 2008. Web.
3. "The Evolution of the Internet: From Military Experiment to General Purpose Technology." *Taylor & Francis*. 8 May 2016. Web.
4. *Four Thieves Vinegar*. Web.
5. Gelles, David. "Tech Backlash Grows as Investors Press Apple to Act on Children's Use." The New York Times. The New York Times, 09 Jan. 2018. Web.
6. "BlackRock's Message: Contribute to Society, or Risk Losing Our Support." *The New York Times*. The New York Times, 16 Jan. 2018. Web.
7. Kim, Larry. "RIP, Facebook News Feed for Publishers." *Inc.com*. Inc., 11 Jan. 2018. Web.
8. Alter, Adam. "Why Our Screens Make Us Less Happy." *TED: Ideas worth Spreading*. 14 Jul. 2017. Web.

14. Secure the World

1. Andrews, Evan. "Who Invented the Internet?" *History.com*. A&E Television Networks, 18 Dec. 2013. Web.
2. "VNI Global Fixed and Mobile Internet Traffic Forecasts." *Cisco*. 27 Nov. 2018. Web.
3. Mcleod, Saul. "Maslow's Hierarchy of Needs." Simply Psychology. Simply Psychology, 21 May 2018. Web.